String Theory

String Theory

By

Dr. Jon Schiller, PhD

String Theory

All rights reserved. No part of this book may be reproduced in any form without written permission from the author, except by a reviewer who may quote brief passages in a review to be printed in a newspaper, a magazine or a website.

WRITTEN BY JON SCHILLER, PhD
Printed in the United States of America
2016
First Printing

ISBN-13: 9781532881671
JON SCHILLER SOFTWARE
jonsch1@verizon.net
http://www.jonschilleroptions.com/
© 2016 by Emilie M. Smyth

String Theory

By

Dr. Jon Schiller, PhD

2016

String Theory

Dr. Jon Schiller's books

Options Trading Books
Insider's Automatic Options Strategy
Self-Adaptive Options & Currency Trading
The 100% Return Options Trading Strategy
Avoid Financial Fraud by Using Weekly Options
Weekly Options Trading to Maximize Your Capital
Options Profits Using Decision Charts Using Strategies
 Developed over 2 Decades of Options Trading
Weekly OEX Options to Grow Trading Capital Rapidly
OEX Weekly Trading Tips & Newsletter Compilations
Double Your Money with Weekly Options Condors
Grow Your Wealth Using Weekly Options Trading
9 Weeks of Trading Weekly Options
Options Strategy to Profit during Extreme Volatility
Options Profits using Sustainable Energy Companies
Weekly Index Options Trading Tips to Increase Profits
Weekly Options Profits Using Tablet Computer
Trade Weekly Options Using Android Mobile Devices
Profit when Algorithmic Trading Systems Cause
 Flashcrashes
Mobile Devices Revolution
Weekly Index Options Impacted by Washington Programs
Weekly Options Trading Algorithm using 2sig & WWI
Win with Weekly Options
Weekly Options in 2015
Weekly Options for Monthly Income
Use Weekly Options to Increase Wealth

Jon Schiller's Documentary & Technical Books
21st Century Cosmology
Quantum Computers,
Life Style to Extend Life Span
Visual Basic Express & Java
Internet View of the Arabic World

String Theory

Bullet Trains Go Over 365 mph
Global Change & Energy Policy
Human Evolution: Neanderthals & Homosapiens
Nano Technology Developments
Big Bang and Black Holes
Avoid Terrorist Attacks
Cyber Attacks & Protection
US Government Debt Story
Education in the 21st Century
Prostate Cancer
Chemical Bonding
Genome Mapping
Bin Laden Demise as Seen on Internet in 2011
Making Movies using PowerPoint, CamStudio,
    WinMovieMaker & WinDVDMaker
Will Barak Obama Win 2012 Presidential Elections?
Wikipedia & Free Speech
History of Civilization
Walking to Improve Health
Great Technology Transitions of Civilization
Higgs Boson Particle and Impact on Cosmology
Mars Mission and Shuttle History & Replacement
Political Finger Pointing Impact on Wall Street
MidEast Revolutions, Iraq, Afghanistan, Egypt, Libya,
    Syria, Tunisia, Yemen, Israel vs Enemies.
Nanorobots & Microrobots   Exciting Tools of Future
GPS Technology for Walking, Driving, Boating, Flying
DNA Chemistry: DNA Damage & Repair; Aid to Human
    Health
Ruby on Rails for Creating Interactive Websites
Dreamliner
Dark Matter Evidence found at CERN
Optogenetics Research
Washington Political Battles
Global Education for Universities and Colleges
Spy Whistleblower

String Theory

Intelligent Drones Future for Military Aircraft
Fly by Wire Aircraft, Fighters, Drones, and Airliners
How to Grow Old Gracefully
Wealth Accumulating to the Few
Revising US Constitution to make it Suitable for Now
Open Heart Bypass Surgery to Fix Defects
F-35 Lightning II
Russian Cultural Tour-Saint Petersburg and Moscow
Anti-Aging Research   Live Much Longer
Aerospace Wonder   Stealth Aircraft
Brain Study 2015
Private Flying in Small Aircraft
How to Eliminate ISIS
Around the World Trip in a LearJet

Fiction Novels & True Stories
IBEX
Lost in Space
Masada Never Again
Multihulls
Ultra Taiwan Fighter
Irrational Indictment & Imprisonment
Family History of a Successful Aerospace Executive
Six Gay Love Tales
Six Gay Love Tales, Vol.2
Five Gay Love Conversions
Gay Love Techniques

String Theory

Table of Contents

Introduction			ix
Chapter	1	Richard Feynman	1
Chapter	2	Definition of String Theory	5
Chapter	3	Black Holes	25
Chapter	4	Connections to Mathematics	41
Chapter	5	History of String Theory	49
Chapter	6	Beyond Nanotech to Femtotech	67
Chapter	7	Femtotechnology 2016	89
Chapter	8	Femtotech (Sub) Nuclear Scale Engineering & Computation	93
Chapter	9	Cosmological Implications of String Theory	109
Chapter	10	What are Strings *made of* in String Theory?	113
Appendix A		References & Bibliography	117

String Theory

String Theory

Introduction

Several time each year I receive *Engineering & Science,* a magazine published at Caltech where I was awarded the BS in Physics in 1951. I learn all of the new areas of research being done at my alma mater. I am especially interested in Physics. This year I learned that research is taking place in String Theory. In physics, *string theory* is a theoretical framework in which the point-like particles of Particle Physics are replaced by one dimensional objects called strings. It describes how these strings propagate through space and interact with each other. On distance scales larger than the string scale, a string looks just like an ordinary particle, with its mass, charge, and other properties determined by the vibrational state of the string. In string theory, one of the many vibrational states of the string corresponds to the graviton, a quantum mechanical particle that carries gravitational force. Thus string theory is a theory of quantum gravity.

String theory is a broad and varied subject that attempts to address a number of deep questions of fundamental physics. String theory has been applied to a variety of problems in black hole physics, early universe cosmology, nuclear physics, and condensed matter physics, and it has stimulated a number of major developments in pure mathematics. Because string theory potentially provides a unified description of gravity and particle physics, it is a candidate for a theory of everything, a self-contained mathematical model

String Theory

that describes all fundamental forces and forms of matter. Despite much work on these problems, it is not known to what extent string theory describes the real world or how much freedom the theory allows to choose the details.

One of the challenges of string theory is that the full theory does not yet have a satisfactory definition in all circumstances. Another issue is that the theory is thought to describe an enormous landscape of possible universes, and this has complicated efforts to develop theories of particle physics based on string theory. These issues have led some in the community to criticize these approaches to physics and question the value of continued research on string theory unification.

String Theory

Chapter 1. Richard Feynman

The nanotech field was arguably launched by Richard Feynman's 1959 talk "There's Plenty of Room at the Bottom." As Feynman wrote there, *"It is a staggeringly small world that is below. In the year 2000, when they look back at this age, they will wonder why it was not until the year 1960 that anybody began seriously to move in this direction.*

Why can't we write the entire 24 volumes of the Encyclopedia Britannica on the head of a pin?" Feynman queried in 1959. He offered $1000 to the first person who could reduce the page of a book to 1/25,000 linear scale, readable by an electron microscope. He also offered the same amount to the first person to build a miniature motor no bigger than $1/6^{th}$ inch cube.

He had to pay out on both prizes – the first within a year of the challenge to Bill McLellan, an electrical engineer and CalTech alumnus. Feynman knew that McLellen was serious when he brought a microscope with him to show Feynman his miniature motor capable of generating a millionth of a horsepower. Although Feynman paid McLellen the prize money in 1960, he was disappointed in the motor because it did not require any technological

String Theory

advances. In an updated version of his talk given twenty years later, Feynman speculated that, with modern technology, it should be possible to mass-produce motors $1/40^{th}$ of a side smaller than McLellen's original motor. To produce such micromachines, Feynman envisioned the cretion of a chain of "slave" machines, each producing tools and machines at one-fourth their own scale.

(Above) The diagrams in the background of this picture were added much later in order to market Feynman's lecture – Feynman articulated the core nanotech vision but he didn't use the word "nanotech" nor depict details of this nature.

In 1985 Feynman had to pay out on the second prize. The winner was Tom Newman, a Stanford graduate student who was using electron beam lithography to engrave patterns on

String Theory

silicon to make integrated circuits. He calculated he would have to reduce individual letters down to a scale only 50 atoms wide. To check that the prize was still being offered he sent a telegram to Feynman. He was surprised to receive a telephone call from Feynman confirming that it was. Newman programed an electron beam machine to write the first page of Charles Dicken's novel *A Tale of Two Cities*. The major difficulty turned out to be actually finding the tiny page on the surface after it had been written. Newman duly received a check from Feynman on November 1985.

Your Author took Advanced Physics from Professor Feynman while I was earning my BS in Physics from Caltech in 1951.

I had no inkling at the time that what he was teaching me would result in Nanotechnology Chips which would replace warehouse sized 90k IBM Computers (implemented with vacuum tubes) with Desktop Computers and Smartphones (implemented with Nano chips). These new computers and smartphones would be available to a large segment of the population in 2001 (and later) at low cost.

Professor Feynman also predicted Femtotechnology. I predict that Femtotechnology will result in a greater revolution in electronics and computational power than Nanotechnology permitted. By April 2016 we have seen no results to the general public of Femtotechnology electronics, but the accelerating pace of Technology will permit

String Theory

Femtotechnology to happen more rapidly than Nanotechnology.

35 Years after Professor Feynman's physics course, your Author designed the Avionics for the F111 fighter jet which used 2 twin IBM airborne computers for the weapons computations. These same twin IBM computers were used on the Space Shuttle with the F111 computer software adapted to the Space Shuttle These computers used microchips rather than vacuum tubes.

Chapter 2. Definition of String Theory

Fundamentals Using Probability

(Above) The fundamental objects of string theory are open and closed strings.

In the twentieth century, two theoretical frameworks emerged for formulating the laws of physics. One of these frameworks was Albert Einstein's General Theory of Relativity, a theory that explains the force of gravity and the structure of space and time. The other was Quantum Mechanics, a radically different formalism for describing physical phenomena using probability. By the late 1970s, these two frameworks had proven to be sufficient to explain most of the observed features of the universe, from elementary particles to atoms to the evolution of stars and the universe as a whole.

String Theory

In spite of these successes, there are still many problems that remain to be solved. One of the deepest problems in modern physics is the problem of Quantum Gravity. The general theory of relativity is formulated within the framework of classical physics, whereas the other fundamental forces are described within the framework of quantum mechanics. A quantum theory of gravity is needed in order to reconcile general relativity with the principles of quantum mechanics, but difficulties arise when one attempts to apply the usual prescriptions of quantum theory to the force of gravity. In addition to the problem of developing a consistent theory of quantum gravity, there are many other fundamental problems in the physics of atomic nuclei, black holes, and the early universe.

String theory is a theoretical framework that attempts to address these questions and many others. The starting point for string theory is the idea that the point-like particles of Particle Physics can also be modeled as one-dimensional objects called strings. String theory describes how strings propagate through space and interact with each other. In a given version of string theory, there is only one kind of string, which may look like a small loop or segment of ordinary string, and it can vibrate in different ways. On distance scales larger than the string scale, a string will look just like an ordinary particle, with its mass, charge, and other properties determined by the vibrational state of the string. In this way, all of the different elementary particles may be viewed as vibrating strings. In string theory, one of the vibrational states of the string gives rise to the graviton, a

String Theory

quantum mechanical particle that carries gravitational force. Thus string theory is a theory of quantum gravity.

One of the main developments of the past several decades in string theory was the discovery of certain "dualities", mathematical transformations that identify one physical theory with another. Physicists studying string theory have discovered a number of these dualities between different versions of string theory, and this has led to the conjecture that all consistent versions of string theory are subsumed in a single framework known as M-theory.

Studies of string theory have also yielded a number of results on the nature of black holes and their gravitational interaction. There are certain paradoxes that arise when one attempts to understand the quantum aspects of black holes, and work on string theory has attempted to clarify these issues. In late 1997 this line of work culminated in the discovery of the anti-de Sitter/conformal field theory correspondence or AdS/CFT. This is a theoretical result which relates string theory to other physical theories which are better understood theoretically. The AdS/CFT correspondence has implications for the study of black holes and quantum gravity, and it has been applied to other subjects, including Nuclear and Condensed Matter Physics.

Since string theory incorporates all of the fundamental interactions, including gravity, many physicists hope that it fully describes our universe, making it a **Theory of Everything**. One of the goals of current research in string

String Theory

theory is to find a solution of the theory that reproduces the observed spectrum of elementary particles, with a small cosmological constant, containing dark matter and a plausible mechanism for cosmic inflation. While there has been progress toward these goals, it is not known to what extent string theory describes the real world or how much freedom the theory allows to choose the details.

Strings

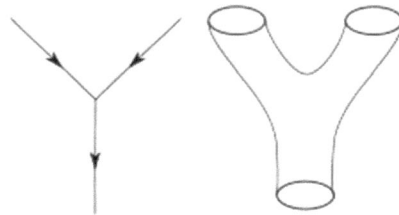

(Above) Interaction in the quantum world: worldliness of point-like particles or a world sheet swept up by closed strings in string theory.

One of the challenges of string theory is that the full theory does not yet have a satisfactory definition in all circumstances. The scattering of strings is most straightforwardly defined using the techniques of perturbation theory, but it is not known in general how to define string theory non-perturbatively. It is also not clear whether there is any principle by which string theory selects its vacuum state, the physical state that determines the properties of our universe. These problems have led some

String Theory

in the community to criticize these approaches to the unification of physics and question the value of continued research on these problems.

The application of quantum mechanics to physical objects such as the electromagnetic field, which are extended in space and time, is known as Quantum Field Theory. In particle physics, quantum field theories form the basis for our understanding of elementary particles, which are modeled as excitations in the fundamental fields.

In quantum field theory, one typically computes the probabilities of various physical events using the techniques of perturbation theory. Developed by Richard Feynman and others in the first half of the twentieth century, perturbative quantum field theory uses special diagrams called Feynman Diagrams to organize computations. One imagines that these diagrams depict the paths of point-like particles and their interactions.

The starting point for string theory is the idea that the point-like particles of quantum field theory can also be modeled as one-dimensional objects called strings. The interaction of strings is most straightforwardly defined by generalizing the perturbation theory used in ordinary quantum field theory. At the level of Feynman Diagrams, this means replacing the one-dimensional diagram representing the path of a point particle by a two-dimensional surface representing the motion of a string. Unlike in quantum field theory, string theory does not yet have a full non-perturbative definition.

String Theory

Thus many of the theoretical questions that physicists would like to answer remain out of reach.

In theories of particle physics based on string theory, the characteristic length scale of strings is assumed to be on the order of the Planck length, or 10^{-35} meters, the scale at which the effects of quantum gravity are believed to become significant. On much larger length scales, such as the scales visible in physics laboratories, such objects would be indistinguishable from zero-dimensional point particles, and the vibrational state of the string would determine the type of particle. One of the vibrational states of a string corresponds to the graviton, a quantum mechanical particle that carries the gravitational force.

The original version of string theory was bosonic string theory, but this version described only bosons, a class of particles which transmit forces between the matter particles, or fermions. Bosonic string theory was eventually superseded by theories called superstring theories. These theories describe both bosons and fermions, and they incorporate a theoretical idea called supersymmetry. This is a mathematical relationship that exists in certain physical theories between the bosons and fermions. In theories with supersymmetry, each boson has a counterpart which is a fermion, and vice versa.

There are several versions of superstring theory: type I, type IIA, type IIB, and two flavors (see next paragraph) of heterotic string theory ($SO(32)$ and $E_8 \mathrm{x} E_8$). The different

String Theory

theories allow different types of strings, and the particles that arise at low energies exhibit different symmetries. For example, the type I theory includes both open strings (which are segments with endpoints) and closed strings (which form closed loops), while types IIA and IIB include only closed strings.

Flavor is the name scientists give to different versions of the same type of particle. For instance, quarks (which make up the protons and neutrons inside atoms) come in six flavors: up, down, top, bottom, strange and charm. Particles called leptons, a category that includes electrons, also come in six flavors, each with a different mass. *This paragraph was written by Clara Moskowitz on 26 January 2012 and found on the Internet.*

Extra Dimension

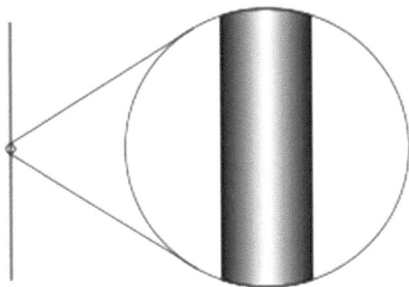

(Above) An example of compactification: At large distances, a two dimensional surface with one circular dimension looks one dimensional.

String Theory

In everyday life, there are three familiar dimensions of space: height, width and length. Einstein's General Theory of Relativity treats Time as a dimension on par with the three spatial dimensions; in General Relativity, space and time are not modeled as separate entities but are instead unified to a four-dimensional space-time. In this framework, the phenomenon of gravity is viewed as a consequence of the geometry of space-time.

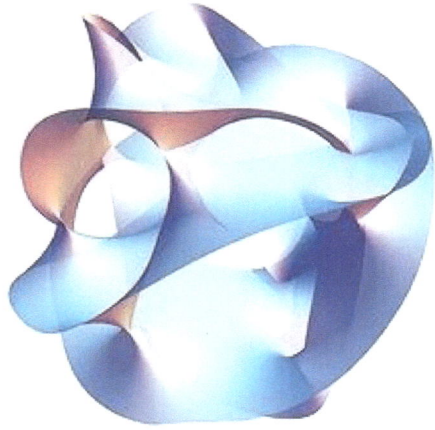

(Above) A Cross section of a Quintic Calabi-Yau Manifold.

There are several reasons why physicists consider theories in other dimensions (in spite of the fact that the universe is well described by four-dimensional space-time). In some cases, by modeling space-time in a different number of dimensions, a theory becomes more mathematically tractable, and one can perform calculations and gain general insights more easily. There are also situations where theories in two or three space-time dimensions are useful for

String Theory

describing phenomena in Condensed Matter Physics. Finally, there exist scenarios in which there could actually be more than four dimensions of space-time which have nonetheless managed to escape detection.

One notable feature of string theories is that these theories require extra dimensions of space-time for their mathematical consistency. In bosonic string theory, space-time is **26**-dimensional, while in superstring theory it is **ten**-dimensional. In order to describe real physical phenomena using string theory, one must therefore imagine scenarios in which these extra dimensions would not be observed in experiments.

Compactification is one way of modifying the number of dimensions in a physical theory. In compactification, some of the extra dimensions are assumed to "close up" on themselves to form circles. In the limit where these curled up dimensions become very small, one obtains a theory in which space-time has effectively a lower number of dimensions. A standard analogy for this is to consider a multi-dimensional object such as a garden hose. If the hose is viewed from a sufficient distance, it appears to have only one dimension, its length. However, as one approaches the hose, one discovers that it contains a second dimension, its circumference. Thus, an ant crawling on the surface of the hose would move in two dimensions.

Compactification can be used to construct models in which space-time is effectively four-dimensional. However, not

String Theory

every method of compactifying the extra dimensions has produced a model with the right properties to describe nature. In a viable model of particle physics, the compact extra dimensions must be shaped like a Calabi-Yau manifold (See page 12). A Calabi-Yau manifold is a special space which is typically taken to be six-dimensional in applications to string theory. It is named after mathematicians Eugenio Calabi and Shing-Tung Yau.

Another approach to reducing the number of dimensions is the so-called Brane-World scenario. In this approach, physicists assume that the observable universe is a four-dimensional subspace of a higher dimensional space. In such models, the force-carrying bosons of particle physics arise from open strings with endpoints attached to the four-dimensional subspace, while gravity arises from closed strings propagating through the larger ambient space. This idea plays an important role in attempts to develop models of real world physics based on string theory, and it provides a natural explanation for the weakness of gravity compared to the other fundamental forces.

Dualities

One notable fact about string theory is that the different versions of the theory all turn out to be related in highly non-trivial ways. One of the relationships that can exist between different string theories is called S-duality. This is a relationship which says that a collection of strongly interacting particles in one theory can, in some cases, be

String Theory

viewed as a collection of weakly interacting particles in a completely different theory. Roughly speaking, a collection of particles is said to be strongly interacting if they combine and decay often and weakly interacting if they combine and decay infrequently. Type I string theory turns out to be equivalent by S-duality to the $SO(32)$ heterotic string theory. Similarly, type IIB string theory is related to itself in a non-trivial way by S-duality. (See S-duality on page 21.)

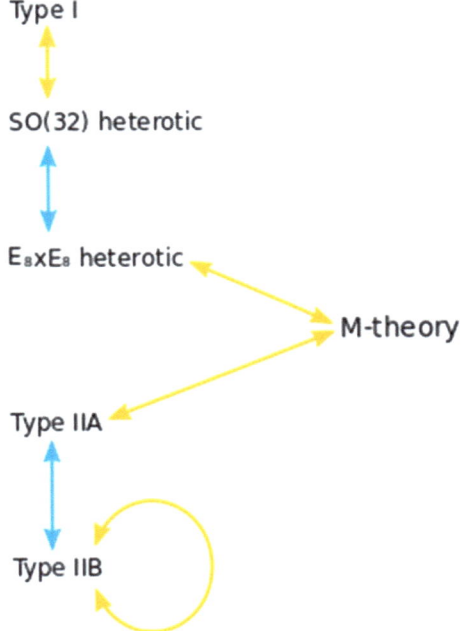

(Above) A diagram of string theory dualities. Yellow arrows indicate S-duality. Blue arrows indicate T-duality.

String Theory

Another relationship between different string theories is T-duality. Here one considers strings propagating around a circular extra dimension. T-duality states that a string propagating around a circle of radius R is equivalent to a string propagating around a circle of radius $1/R$ in the sense that all observable quantities in one description are identified with quantities in the dual description. For example, a string has momentum as it propagates around a circle, and it can also wind around the circle one or more times. The number of times the string winds around a circle is called the winding number. If a string has momentum p and winding number n in one description, it will have momentum n and winding number p in the dual description. For example, type IIA string theory is equivalent to type IIB string theory via T-duality, and the two versions of heterotic string theory are also related by T-duality.

In general, the term *duality* refers to a situation where two seemingly different physical systems turn out to be equivalent in a non-trivial way. Two theories related by a duality need not be string theories. For example, Montonen-Olive duality is an example of an S-duality relationship between quantum field theories. The AdS/CFT correspondence is an example of a duality which relates string theory to a quantum field theory. If two theories are related by a duality, it means that one theory can be transformed in some way so that it ends up looking just like the other theory. The two theories are then said to be *dual* to one another under the transformation. Put differently, the

String Theory

two theories are mathematically different descriptions of the same phenomena.

Branes

In string theory and related theories, a brane is a physical object that generalizes the idea of a point particle to higher dimensions. For example, a point particle can be viewed as a brane of dimension zero, while a string can be viewed as a brane of dimension one. It is also possible to consider higher-dimensional branes. In dimension p, these are called p-branes. The word brane comes from the word "membrane" which refers to a two-dimensional brane.

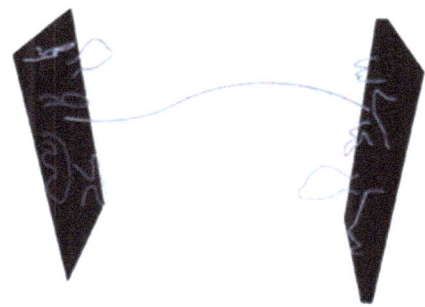

(Above) Open strings attached to a pair of D-branes.

Branes are dynamic objects which can propagate through space-time according to the rules of quantum mechanics. They have mass and can have other attributes such as charge. A p-brane sweeps out a $(p+1)$-dimensional volume in space-time called its *world-volume*. Physicists often study fields analogous to the electromagnetic field which live on the world-volume of a brane.

String Theory

In string theory, D-branes are an important class of branes that arise when one considers open strings. As an open string propagates through space-time, its endpoints are required to lie on a D-brane. The letter "D" in D-brane refers to a certain mathematical condition on the system known as the Dirichlet boundary condition. The study of D-branes in string theory has led to important results such as the AdS/CFT correspondence, which has shed light on many problems in quantum field theory.

Branes are also frequently studied from a purely mathematical point of view. Mathematically, branes can be described as objects of certain categories, such as the derived category of coherent sheaves on a complex algebraic variety, or the Fukaya category of a symplectic manifold. The connection between the physical notion of a brane and the mathematical notion of a category has led to important mathematical insights in the fields of algebraic and simplectic geometry and representation theory.

M-Theory

Prior to 1995, theorists believed that there were five consistent versions of superstring theory (type I, type IIA, type IIB, and two versions of heterotic string theory). This understanding changed in 1995 when Edward Witten suggested that the five theories were just special limiting cases of an eleven-dimensional theory called M-theory. Witten's conjecture was based on the work of a number of other physicists, including Ashoke Sen, Chris Hull, Paul

String Theory

Townsend, and Michael Duff. His announcement led to a flurry of research activity now known as the Second Superstring Revolution.

Unification of Superstring Theories

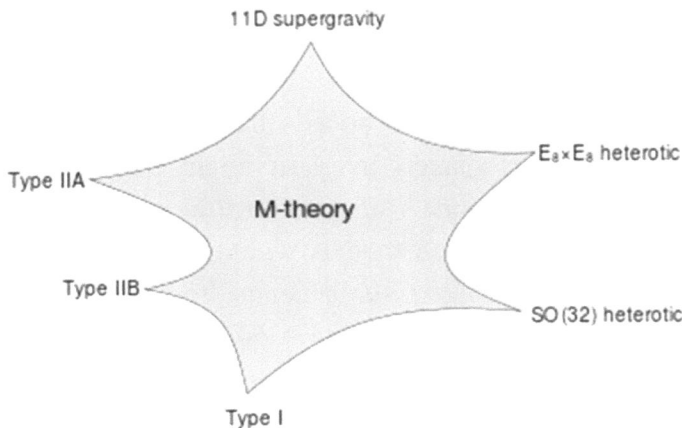

(Above) A schematic illustration of the relationship between M-theory, the five superstring theories, and the eleven-dimensional supergravity. The shaded region represents a family of different physical scenarios that are possible in M-theory. In certain limiting cases corresponding to the cusps, it is natural to describe the physics using one of the six theories labeled there.

In the 1970s, many physicists became interested in supergravity theories, which combine general relativity with supersymmetry. Whereas general relativity makes sense in any number of dimensions, supergravity places an upper

String Theory

limit on the number of dimensions. In 1978, work by Werner Nahm showed that the maximum space-time dimension in which one can formulate a consistent supersymmetric theory is eleven. In the same year, Eugene Cremmer, Bernard Julia, and Joel Scherk of the École Normale Supérieure showed that supergravity not only permits up to eleven dimensions but is in fact most elegant in this maximum number of dimensions.

Initially, many physicists hoped that by compactifying eleven-dimensional supergravity, it might be possible to construct realistic models of our four-dimensional world. The hope was that such models would provide a unified description of the four fundamental forces of nature: electromagnetism, the strong and weak nuclear forces, and gravity. Interest in eleven-dimensional supergravity soon waned as various flaws in this scheme were discovered. One of the problems was that the laws of physics appear to distinguish between clockwise and counterclockwise, a phenomenon known as chirality. Edward Witten and others observed this chirality property cannot be readily derived by compactifying from eleven dimensions.

In the first superstring revolution in 1984, many physicists turned to string theory as a unified theory of particle physics and quantum gravity. Unlike supergravity theory, string theory was able to accommodate the chirality of the standard model, and it provided a theory of gravity consistent with quantum effects. Another feature of string theory that many physicists were drawn to in the 1980s and 1990s was its high

String Theory

degree of uniqueness. In ordinary particle theories, one can consider any collection of elementary particles whose classical behavior is described by an arbitrary Lagrangian. In string theory, the possibilities are much more constrained. By the 1990s, physicists argued that there were only five consistent supersymmetric versions of the theory.

Although there were only a handful of consistent superstring theories, it remained a mystery why there was not just one consistent formulation. However, as physicists began to examine string theory more closely, they realized that these theories are related in intricate and nontrivial ways. They found that a system of strongly interacting strings can, in some cases, be viewed as a system of weakly interacting strings. This phenomenon is known as S-duality. It was studied by Ashoke Sen in the context of heterotic strings in four dimensions and by Chris Hull and Paul Townsend in the context of the type IIB theory. Theorists also found that different string theories may be related by T-duality. This duality implies that strings propagating on completely different space-time geometries may be physically equivalent.

At around the same time as many physicists were studying the properties of strings, a small group of physicists was examining the possible applications of higher dimensional objects. In 1987, Eric Bergshoeff, Ergin Sezgin, and Paul Townsend showed that eleven-dimensional supergravity includes two-dimensional branes. Intuitively, these objects look like sheets or membranes propagating through the

String Theory

eleven-dimensional space-time. Shortly after this discovery, Michael Duff, Paul Howe, Takeo Imani, and Kellogg Stelle considered a particular compactification of eleven-dimensional supergravity with one of the dimensions curled up into a circle. In this setting, one can imagine the membrane wrapping around the circular dimension. If the radius of the circle is sufficiently small, then this membrane looks just like a string in ten-dimensional space-time. In fact, Duff and his collaborators showed that this construction reproduces exactly the strings appearing in type IIA superstring theory.

Speaking at a string theory conference in 1995, Edward Witten made the surprising suggestion that all five superstring theories were in fact just different limiting cases of a single theory in eleven space-time dimensions. Witten's announcement drew together all of the previous results on S- and T-duality and the appearance of higher dimensional branes in string theory. In the months following Witten's announcement, hundreds of new papers appeared on the Internet confirming different parts of his proposal. Today this flurry of work is known as the Second Superstring Revolution (mentioned on page 19.).

Initially, some physicists suggested that the new theory was a fundamental theory of membranes, but Witten was skeptical of the role of membranes in the theory. In a paper from 1996, Hořava and Witten wrote "As it has been proposed that the eleven-dimensional theory is a super-membrane theory but there are some reasons to doubt that

String Theory

interpretation, we will non-committedly call it the M-theory, leaving to the future the relation of M to membranes." In the absence of an understanding of the true meaning and structure of M-theory, Witten has suggested that the *"M"* should stand for "magic", "mystery", or "membrane" according to taste, and the true meaning of the title should be decided when a more fundamental formulation of the theory is known.

Matrix Theory Physics

In mathematics, a matrix is a rectangular array of numbers or other data. In physics, a matrix model is a particular kind of physical theory whose mathematical formulation involves the notion of a matrix in an important way. A matrix model describes the behavior of a set of matrices within the framework of quantum mechanics.

One important example of a matrix model is the BFSS matrix model proposed by Tom Banks, Willy Fischler, Stephen Shenker, and Leonard Susskind in 1997. This theory describes the behavior of a set of nine large matrices. In their original paper, these authors showed, among other things, that the low energy limit of this matrix model is described by eleven-dimensional supergravity. These calculations led them to propose that the BFSS matrix model is exactly equivalent to M-theory. The BFSS matrix model can therefore be used as a prototype for a correct formulation of M-theory and a tool for investigating the properties of M-theory in a relatively simple setting.

String Theory

The development of the matrix model formulation of M-theory led physicists to consider various connections between string theory and a branch of mathematics called Noncommutative Geometry. This subject is a generalization of ordinary geometry in which mathematicians define new geometric notions using tools from Noncommutative Algebra. In a paper from 1998, Alain Connes, Michael R. Douglas, and Albert Schwarz showed that some aspects of matrix models and M-theory are described by a noncommutative quantum field theory, a special kind of physical theory in which space-time is described mathematically using noncommutative geometry. This established a link between matrix models and M-theory on the one hand, and noncommutative geometry on the other hand. It quickly led to the discovery of other important links between noncommutative geometry and various physical theories.

String Theory

Chapter 3. Black Holes

In general relativity, a black hole is defined as a region of space-time in which the gravitational field is so strong that no particle or radiation can escape. In the currently accepted models of stellar evolution, black holes are thought to arise when massive stars undergo gravitational collapse, and many galaxies are thought to contain supermassive black holes at their centers. Black holes are also important for theoretical reasons, as they present profound challenges for theorists attempting to understand the quantum aspects of gravity. String theory has proved to be an important tool for investigating the theoretical properties of black holes because it provides a framework in which theorists can study their thermodynamics.

Bekenstein–Hawking Formula

In the branch of physics called Statistical Mechanics, there is a measure of the randomness or disorder of a physical system. This concept was studied in the 1870s by the Austrian physicist Ludwig Boltzmann, who showed that the thermodynamic properties of a gas could be derived from the combined properties of its many constituent molecules. Boltzmann argued that by averaging the behaviors of all the different molecules in a gas, one can understand

String Theory

macroscopic properties such as volume, temperature, and pressure. In addition, this perspective led him to give a precise definition of entropy as the natural logarithm of the number of different states of the molecules (also called *microstates*) that give rise to the same macroscopic features.

In the twentieth century, physicists began to apply the same concepts to black holes. In most systems such as gases, the entropy value increases with the volume. In the 1970s, the physicist Jacob Bekenstein suggested that the entropy of a black hole is instead proportional to the *surface area* of its event horizon, the boundary beyond which matter and radiation is lost to its gravitational attraction. When combined with ideas of the physicist Stephen Hawking, Bekenstein's work yielded a precise formula for the entropy of a black hole. The formula expresses the entropy S as

$$S = \frac{c^3 kA}{4\hbar G}$$

where c is the speed of light, k is Boltzmann's constant, h is the reduced Planck constant, G is Newton's constant, and A is the surface area of the event horizon.

Like any physical system, a black hole has an entropy defined in terms of the number of different microstates that lead to the same macroscopic features. The Bekenstein–Hawking entropy formula gives the expected value of the entropy of a black hole, but by the 1990s, physicists still lacked a derivation of this formula by counting microstates in a theory of quantum gravity. Finding such a derivation of

String Theory

this formula was considered an important test of the viability of any theory of quantum gravity such as string theory.

Derivation within String Theory

In a paper from 1996, Andrew Strominger and Cumrun Vafa showed how to derive the Bekenstein–Hawking formula for certain black holes in string theory. Their calculation was based on the observation that D-branes -- which look like fluctuating membranes when they are weakly interacting -- become dense, massive objects with event horizons when the interactions are strong. In other words, a system of strongly interacting D-branes in string theory is indistinguishable from a black hole. Strominger and Vafa analyzed such D-brane systems and calculated the number of different ways of placing D-branes in space-time so that their combined mass and charge is equal to a given mass and charge for the resulting black hole. Their calculations reproduced the Bekenstein–Hawking formula exactly, including the factor of 1/4. Subsequent work by Strominger, Vafa, and others refined the original calculations and gave the precise values of the "quantum corrections" needed to describe very small black holes.

The black holes that Strominger and Vafa considered in their original work were quite different from real astrophysical black holes. One difference was that Strominger and Vafa considered only extremal black holes in order to make the calculation tractable. These are defined as black holes with the lowest possible mass compatible with a given charge.

String Theory

Strominger and Vafa also restricted their attention to black holes in five-dimensional space-time with unphysical supersymmetry.

Although it was originally developed in this very particular and physically unrealistic context in string theory, the entropy calculation of Strominger and Vafa has led to a qualitative understanding of how black hole entropy can be accounted for in any theory of quantum gravity. Indeed, in 1998, Strominger argued that the original result could be generalized to an arbitrary consistent theory of quantum gravity without relying on strings or supersymmetry. In collaboration with several other authors in 2010, he showed that some results on black hole entropy could be extended to non-extremal astrophysical black holes.

AdS/CFT Correspondence

One approach to formulating string theory and studying its properties is provided by the anti-de Sitter/conformal field theory (AdS/CFT) correspondence: this is a theoretical result which implies that string theory is in some cases equivalent to a quantum field theory. In addition to providing insights into the mathematical structure of string theory, the AdS/CFT correspondence has shed light on many aspects of quantum field theory in regimes where traditional calculation techniques are ineffective. The AdS/CFT correspondence was first proposed by Juan Maldacena in late 1997. Important aspects of the correspondence were elaborated in articles by Steven Gubser, Igor Klebanov,

String Theory

Alexander Markovich Polyakov, and Edward Witten. By 2010, Maldacena's article had over 7000 citations, becoming the most highly cited article in the field of high energy physics.

Overview of the Correspondence

(Above) A tessellation of the hyperbolic plane by triangles and squares.

In the AdS/CFT correspondence, the geometry of space-time is described in terms of a certain vacuum solution of Einstein's equation called anti-de Sitter space. In very elementary terms, anti-de Sitter space is a mathematical model of space-time in which the idea of the distance between points (the metric) is different from the idea of distance in ordinary Euclidean geometry. It is closely related to hyperbolic space, which can be viewed as a disk as illustrated above. This image shows a tessellation of a disk

String Theory

by triangles and squares. One can define the distance between points of this disk in such a way that all the triangles and squares are the same size and the circular outer boundary is infinitely far from any point in the interior.

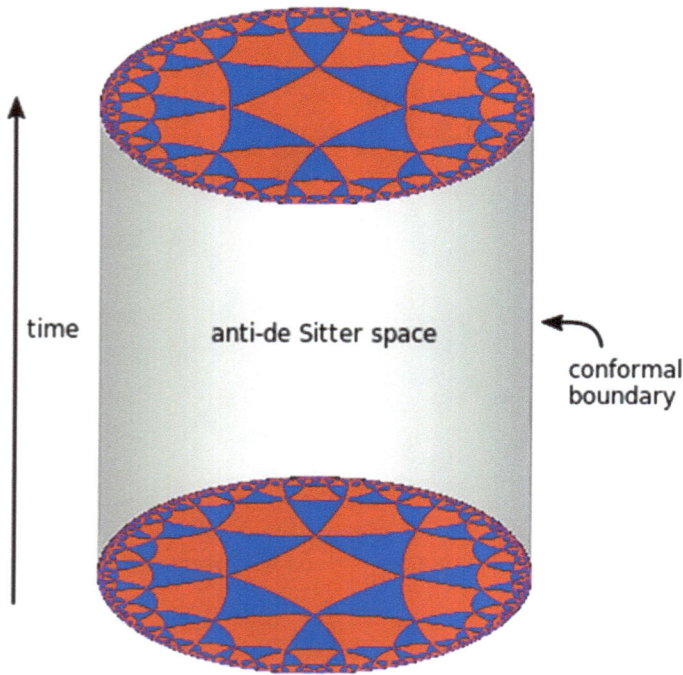

(Above) Three-dimensional anti-de Sitter space is like a stack of hyperbolic disks, each one representing the state of the universe at a given time.

The resulting space-time looks like a solid cylinder in which any cross section is a copy of the hyperbolic disk. Time runs along the vertical direction in this picture. The surface of this cylinder plays an important role in the AdS/CFT

String Theory

correspondence. As with the hyperbolic plane, anti-de Sitter space is curved in such a way that any point in the interior is actually infinitely far from this boundary surface.

This construction describes a hypothetical universe with only two space dimensions and a single time dimension, but it can be generalized to any number of dimensions. Indeed, hyperbolic space can have more than two dimensions and one can "stack up" copies of hyperbolic space to get higher-dimensional models of anti-de Sitter space.

An important feature of anti-de Sitter space is its boundary (which looks like a cylinder in the case of three-dimensional anti-de Sitter space). One property of this boundary is that, within a small region on the surface around any given point, it looks just like Minkowski space, the model of space-time used in non-gravitational physics. One can therefore consider an auxiliary theory in which "space-time" is given by the boundary of anti-de Sitter space. This observation is the starting point for AdS/CFT correspondence, which states that the boundary of anti-de Sitter space can be regarded as the "space-time" for a quantum field theory. The claim is that this quantum field theory is equivalent to a gravitational theory, such as string theory, in the bulk anti-de Sitter space in the sense that there is a "dictionary" for translating entities and calculations in one theory into their counterparts in the other theory. For example, a single particle in the gravitational theory might correspond to some collection of particles in the boundary theory. In addition, the predictions in the two theories are quantitatively identical so that if two

String Theory

particles have a 40 percent chance of colliding in the gravitational theory, then the corresponding collections in the boundary theory would also have a 40 percent chance of colliding.

Applications to Quantum Gravity

The discovery of the AdS/CFT correspondence was a major advance in physicists' understanding of string theory and quantum gravity. One reason for this is that the correspondence provides a formulation of string theory in terms of quantum field theory, which is well understood by comparison. Another reason is that it provides a general framework in which physicists can study and attempt to resolve the paradoxes of black holes.

In 1975, Stephen Hawking published a calculation which suggested that black holes are not completely black but emit a dim radiation due to quantum effects near the event horizon. At first, Hawking's result posed a problem for theorists because it suggested that black holes destroy information. More precisely, Hawking's calculation seemed to conflict with one of the basic postulates of quantum mechanics, which states that physical systems evolve in time according to the Schrödinger equation. This property is usually referred to as unitarity of time evolution. The apparent contradiction between Hawking's calculation and the unitarity postulate of quantum mechanics came to be known as the black hole information paradox.

String Theory

The AdS/CFT correspondence resolves the black hole information paradox, at least to some extent, because it shows how a black hole can evolve in a manner consistent with quantum mechanics in some contexts. Indeed, one can consider black holes in the context of the AdS/CFT correspondence, and any such black hole corresponds to a configuration of particles on the boundary of anti-de Sitter space. These particles obey the usual rules of quantum mechanics and in particular evolve in a unitary fashion, so the black hole must also evolve in a unitary fashion, respecting the principles of quantum mechanics. In 2005, Hawking announced that the paradox had been settled in favor of information conservation by the AdS/CFT correspondence, and he suggested a concrete mechanism by which black holes might preserve information.

Applications to Quantum Field Theory

In addition to its applications to theoretical problems in quantum gravity, the AdS/CFT correspondence has been applied to a variety of problems in quantum field theory. One physical system that has been studied using the AdS/CFT correspondence is the quark–gluon plasma, an exotic state of matter produced in particle accelerators. This state of matter arises for brief instants when heavy ions such as gold or lead nuclei are collided at high energies. Such collisions cause the quarks that make up atomic nuclei to deconfine at temperatures of approximately two trillion kelvins, conditions similar to those present at around

String Theory

10^{-11} seconds after the Big Bang. (*Deconfine* means to free or to be freed from quantum confinement.)

(Above) A magnet levitating above a high temperature superconductor. Today some physicists are working to understand high-temperature superconductivity using the AdS/CFT correspondence.

The physics of the quark–gluon plasma is governed by a theory called quantum chromodynamics, but this theory is mathematically intractable in problems involving the quark–gluon plasma. In an article appearing in 2005, Dàm Thanh Son and his collaborators showed that the AdS/CFT correspondence could be used to understand some aspects of the quark–gluon plasma by describing it in the language of string theory. By applying the AdS/CFT correspondence, Son and his collaborators were able to describe the quark-gluon plasma in terms of black holes in five-dimensional space-time. The calculation showed that the ratio of two quantities associated with the quark–gluon plasma, the shear viscosity and volume density of entropy, should be approximately equal to a certain universal constant. In 2008,

String Theory

the predicted value of this ratio for the quark–gluon plasma was confirmed at the Relativistic Heavy Ion Collider at Brookhaven National Laboratory.

The AdS/CFT correspondence has also been used to study aspects of condensed matter physics. Over the decades, experimental condensed matter physicists have discovered a number of exotic states of matter, including superconductors and superfluids. These states are described using the formalism of quantum field theory, but some phenomena are difficult to explain using standard field theoretic techniques. Some condensed matter theorists including Subir Sachdev hope that the AdS/CFT correspondence will make it possible to describe these systems in the language of string theory and learn more about their behavior.

So far some success has been achieved in using string theory methods to describe the transition of a superfluid to an insulator. A superfluid is a system of electrically neutral atoms that flows without any friction. Such systems are often produced in the laboratory using liquid helium, but recently experimentalists have developed new ways of producing artificial superfluids by pouring trillions of cold atoms into a lattice of crisscrossing lasers. These atoms initially behave as a superfluid, but as experimentalists increase the intensity of the lasers, they become less mobile and then suddenly transition to an insulating state. During the transition, the atoms behave in an unusual way. For example, the atoms slow to a halt at a rate that depends on the temperature and on Planck's constant, the fundamental parameter of quantum mechanics, which does not enter into the description of the other phases. This behavior has recently been understood by considering a dual description where properties of the fluid are described in terms of a higher dimensional black hole.

Phenomenology

In addition to being an idea of considerable theoretical interest, string theory provides a framework for constructing models of real world physics that combine general relativity and particle physics. Phenomenology is the branch of theoretical physics in which physicists construct realistic models of nature from more abstract theoretical ideas. String phenomenology is the part of string theory that attempts to construct realistic models based on string theory.

Partly because of theoretical and mathematical difficulties and partly because of the extremely high energies needed to test these theories experimentally, there is so far no experimental evidence that would unambiguously point to any of these models being a correct fundamental description of nature. This has led some in the community to criticize these approaches to unification and question the value of continued research on these problems.

Particle Physics

The currently accepted theory describing elementary particles and their interactions is known as the standard

String Theory

model of particle physics. This theory provides a unified description of three of the fundamental forces of nature: electromagnetism and the strong and weak nuclear forces. Despite its remarkable success in explaining a wide range of physical phenomena, the standard model cannot be a complete description of reality. This is because the standard model fails to incorporate the force of gravity and because of problems such as the hierarchy problem and the inability to explain the structure of fermion masses or dark matter.

String theory has been used to construct a variety of models of particle physics going beyond the standard model. Typically, such models are based on the idea of compactification. Starting with the ten- or eleven-dimensional space-time of string or M-theory, physicists postulate a shape for the extra dimensions. By choosing this shape appropriately, they can construct models roughly similar to the standard model of particle physics, together with additional undiscovered particles. One popular way of deriving realistic physics from string theory is to start with the heterotic theory in ten dimensions and assume that the six extra dimensions of space-time are shaped like a six-dimensional Calabi–Yau manifold. Such compactifications offer many ways of extracting realistic physics from string theory. Other similar methods can be used to construct realistic models of our four-dimensional world based on M-theory.

String Theory

Cosmology

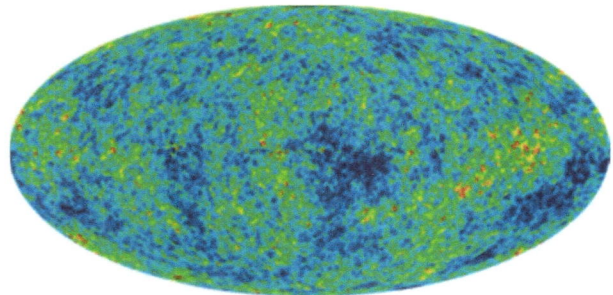

(Above) A map of the cosmic microwave background produced by the Wilkinson Microwave Anisotropy Probe.

The Big Bang theory is the prevailing cosmological model for the universe from the earliest known periods through its subsequent large-scale evolution. Despite its success in explaining many observed features of the universe including galactic redshifts, the relative abundance of light elements such as hydrogen and helium, and the existence of a cosmic microwave background, there are several questions that remain unanswered. For example, the standard Big Bang model does not explain why the universe appears to be the same in all directions, why it appears flat on very large distance scales, or why certain hypothesized particles such as magnetic monopoles are not observed in experiments.

Currently, the leading candidate for a theory going beyond the Big Bang is the theory of cosmic inflation. Developed by Alan Guth and others in the 1980s, inflation postulates a

String Theory

period of extremely rapid accelerated expansion of the universe prior to the expansion described by the standard Big Bang theory. The theory of cosmic inflation preserves the successes of the Big Bang while providing a natural explanation for some of the mysterious features of the universe. The theory has also received striking support from observations of the cosmic microwave background, the radiation that has filled the sky since around 380,000 years after the Big Bang.

In the theory of inflation, the rapid initial expansion of the universe is caused by a hypothetical particle called *the inflation*. The exact properties of this particle are not fixed by the theory but should ultimately be derived from a more fundamental theory such as string theory. Indeed, there have been a number of attempts to identify an inflation within the spectrum of particles described by string theory and to study inflation using string theory. While these approaches might eventually find support in observational data such as measurements of the cosmic microwave background, the application of string theory to cosmology is still in its early stages.

String Theory

Chapter 4. Connections to Mathematics

In addition to influencing research in theoretical physics, string theory has stimulated a number of major developments in pure mathematics. Like many developing ideas in theoretical physics, string theory does not at present have a mathematically rigorous formulation in which all of its concepts can be defined precisely. As a result, physicists who study string theory are often guided by physical intuition to conjecture relationships between the seemingly different mathematical structures that are used to formalize different parts of the theory. These conjectures are later proved by mathematicians, and in this way, string theory serves as a source of new ideas in pure mathematics.

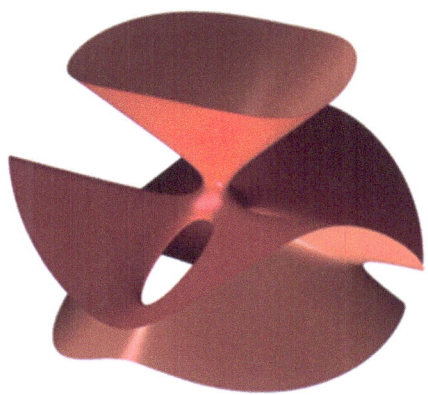

String Theory

The Clebsch cubic (shown on the previous page) is an example of a kind of geometric object called an algebraic variety. A classical result of enumerative geometry states that there are exactly 27 straight lines that lie entirely on this surface.

Mirror Symmetry

After Calabi–Yau manifolds had entered physics as a way to compactify extra dimensions in string theory, many physicists began studying these manifolds. In the late 1980s, several physicists noticed that given such a compactification of string theory, it is not possible to reconstruct uniquely a corresponding Calabi–Yau manifold. Instead, two different versions of string theory, type IIA and type IIB, can be compactified on completely different Calabi–Yau manifolds giving rise to the same physics. In this situation, the manifolds are called Mirror Manifolds, and the relationship between the two physical theories is called mirror symmetry.

Regardless of whether Calabi–Yau compactifications of string theory provide a correct description of nature, the existence of the mirror duality between different string theories has significant mathematical consequences. The Calabi–Yau manifolds used in string theory are of interest in pure mathematics, and mirror symmetry allows mathematicians to solve problems in enumerative geometry, a branch of mathematics concerned with counting the numbers of solutions to geometric questions.

String Theory

Enumerative geometry studies a class of geometric objects called algebraic varieties which are defined by the vanishing of polynomials. For example, the Clebsch cubic illustrated on page 41 is an algebraic variety defined using a certain polynomial of degree three in four variables. A celebrated result of nineteenth-century mathematicians Arthur Cayley and George Salmon states that there are exactly 27 straight lines that lie entirely on such a surface.

Generalizing this problem, one can ask how many lines can be drawn on a quintic Calabi–Yau manifold, such as the one illustrated on page 12, which is defined by a polynomial of degree five. This problem was solved by the nineteenth-century German mathematician Herman Schubert, who found that there are exactly 2,875 such lines. In 1986, geometry expert Sheldon Katz proved that the number of curves, such as circles, that are defined by polynomials of degree two and lie entirely in the quintic is 609,250.

By the year 1991, most of the classical problems of enumerative geometry had been solved and interest in enumerative geometry had begun to diminish. The field was reinvigorated in May 1991 when physicists Philip Candelas, Xenia de la Ossa, Paul Green, and Linda Parks showed that mirror symmetry could be used to translate difficult mathematical questions about one Calabi–Yau manifold into easier questions about its mirror. In particular, they used mirror symmetry to show that a six-dimensional Calabi–Yau manifold can contain exactly 317,206,375 curves of degree three. In addition to counting degree-three curves, Candelas and his collaborators obtained a number of more general results for counting rational curves which went far beyond the results obtained by mathematicians.

Originally, these results of Candelas were justified on physical grounds. However, mathematicians generally prefer rigorous proofs that do not require an appeal to physical intuition. Inspired by physicists' work on mirror symmetry, mathematicians have therefore constructed their own arguments proving the enumerative predictions of mirror symmetry. Today mirror symmetry is an active area of research in mathematics, and mathematicians are working to develop a more complete mathematical understanding of mirror symmetry based on physicists' intuition. Major approaches to mirror symmetry include the homological mirror symmetry program of Maxim Kontsevich and the SYZ conjecture of Andrew Strominger, Shing-Tung Yau, and Eric Zaslow.

Monstrous Moonshine

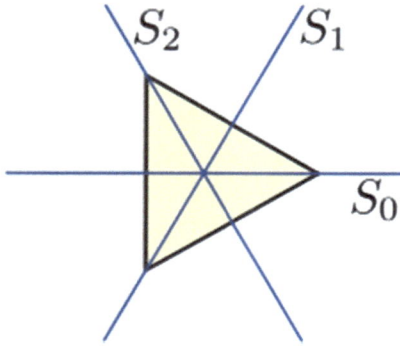

String Theory

(Above) An equilateral triangle can be rotated through 120°, 240°, or 360°, or reflected in any of the three lines pictured without changing its shape.

Group theory is the branch of mathematics that studies the concept of symmetry. For example, one can consider a geometric shape such as an equilateral triangle. There are various operations that one can perform on this triangle without changing its shape. One can rotate it through 120°, 240°, or 360°, or one can reflect in any of the lines labeled S_0, S_1, or S_2 in the picture. Each of these operations is called a *symmetry*, and the collection of these symmetries satisfies certain technical properties making it into what mathematicians call a group. In this particular example, the group is known as the dihedral group of order 6 because it has six elements. A general group may describe finitely many or infinitely many symmetries; if there are only finitely many symmetries, it is called a finite group.

Mathematicians often strive for a classification (or list) of all mathematical objects of a given type. It is generally believed that finite groups are too diverse to admit a useful classification. A more modest but still challenging problem is to classify all finite *simple* groups. These are finite groups which may be used as building blocks for constructing arbitrary finite groups in the same way that prime numbers can be used to construct arbitrary whole numbers by taking products (the answer when multiplying). One of the major achievements of contemporary group theory is the classification of finite simple groups, a mathematical

String Theory

theorem which provides a list of all possible finite simple groups.

This classification theorem identifies several infinite families of groups as well as 26 additional groups which do not fit into any family. The latter groups are called the "sporadic" groups, and each one owes its existence to a remarkable combination of circumstances. The largest sporadic group, the so-called Monster Group, has over 10^{53} elements, more than a thousand times the number of atoms in the Earth.

(Above) A graph of the *J*-Function in the complex plane.

A seemingly unrelated construction is the *j*-function of number theory. This object belongs to a special class of functions called Modular Functions, whose graphs form a certain kind of repeating pattern. Although this function appears in a branch of mathematics which seems very

String Theory

different from the theory of finite groups, the two subjects turn out to be intimately related.

In the late 1970s, mathematicians John McKay and John Thompson noticed that certain numbers arising in the analysis of the Monster Group (namely, the dimensions of its irreducible representations) are related to numbers that appear in a formula for the j-function (namely, the coefficients of its Fourier Series). This relationship was further developed by John Horton Conway and Simon Norton who called it "Monstrous Moonshine" because it seemed so far-fetched.

In 1992, Richard Borcherds constructed a bridge between the theory of Modular Functions and Finite Groups and, in the process, explained the observations of McKay and Thompson. Borcherds' work used ideas from string theory in an essential way, extending earlier results of Igor Frenkel, James Lepowsky, and Arne Meurman, who had realized the Monster Group as the symmetries of a particular version of string theory. In 1998, Borcherds was awarded the Fields Medal for his work.

Since the 1990s, the connection between string theory and moonshine has led to further results in mathematics and physics. In 2010, physicists Tohru Eguchi, Hirosi Ooguri, and Yuji Tachikawa discovered connections between a different sporadic group, the Mathieu Group M_{24}, and a certain version of string theory. Miranda Cheng, John Duncan, and Jeffrey A. Harvey proposed a generalization of this moonshine phenomenon called Umbral Moonshine, and

String Theory

their conjecture was proved mathematically by Duncan, Michael Griffin, and Ken Ono. Witten has also speculated that the version of string theory appearing in monstrous moonshine might be related to a certain simplified model of gravity in three space-time dimensions.

String Theory

Chapter 5. History of String Theory

Early Results

Some of the structures reintroduced by string theory arose for the first time much earlier as part of the program of classical unification started by Albert Einstein. The first person to add a fifth dimension to a theory of gravity was Gunnar Nordström in 1914, who noted that gravity in five dimensions describes both gravity and electromagnetism in four dimensions. Nordström attempted to unify electromagnetism with his theory of gravitation, which was however superseded by Einstein's general relativity in 1919. Thereafter, German mathematician Theodor Kaluza combined the fifth dimension with general relativity, and only Kaluza is usually credited with the idea. In 1926, the Swedish physicist Oskar Klein gave a physical interpretation of the unobservable extra dimension: "It is wrapped into a small circle." Einstein introduced a non-symmetric metric tensor, while much later Brans and Dicke added a scalar component to gravity. These ideas would be revived within string theory, where they are required for consistency.

String Theory

(Above) Leonard Susskind

String theory was originally developed during the late 1960s and early 1970s as a never completely successful theory of hadrons, the subatomic particles like the proton and neutron that feel strong interaction. In the 1960s, Geoffrey Chew and Steven Frautschi discovered that the mesons make families called Regge Trajectories with masses related to spins in a way that was later understood by Yoichiro Nambu, Holger Bech Nielsen and Leonard Susskind to be the relationship expected from rotating strings. Chew advocated making a theory for the interactions of these trajectories that did not presume that they were composed of any fundamental particles, but would construct their interactions from self-consistency conditions on the S-matrix. The S-matrix approach was started by Werner Heisenberg in the 1940s as a way of constructing a theory that did not rely on the local notions of space and time, which Heisenberg believed break

String Theory

down at the nuclear scale. While the scale was off by many orders of magnitude, the approach he advocated was ideally suited for a theory of quantum gravity.

Working with experimental data, R. Dolen, D. Horn and C. Schmid developed some sum rules for hadron exchange. When a particle and an antiparticle scatter, virtual particles can be exchanged in two qualitatively different ways. In the s-channel, the two particles annihilate to make temporary intermediate states that fall apart into the final state particles. In the t-channel, the particles exchange intermediate states by emission and absorption. In field theory, the two contributions add together, one giving a continuous background contribution, the other giving peaks at certain energies. In the data, it was clear that the peaks were stealing from the background -- the authors interpreted this as meaning that the t-channel contribution was dual to the s- channel one, meaning both described the whole amplitude and included the other.

The result was widely advertised by Murray Gell-Mann, leading Gabriele Veneziano to construct a scattering amplitude that had the property of Dolen-Horn-Schmid duality, later renamed "world-sheet duality." The amplitude needed poles where the particles appear, on straight line trajectories, and there is a special mathematical function whose poles are evenly spaced on half the real line – the Gamma Function -- which was widely used in Regge Theory. By manipulating combinations of Gamma functions, Veneziano was able to find a consistent scattering

String Theory

amplitude with poles on straight lines, with mostly positive residues, which obeyed duality and had the appropriate Regge scaling at high energy. The amplitude could fit near-beam scattering data as well as other Regge type fits, and had a suggestive integral representation that could be used for generalization.

(Above) Gabriele Veneziano

Over the next years, hundreds of physicists worked to complete the bootstrap program for this model. Many surprises came up. Veneziano himself discovered that for the scattering amplitude to describe the scattering of a particle that appeared in the theory, an obvious self-consistency condition, the lightest particle had to be a tachyon. Miguel Virasoro and Joel Shapiro found a different amplitude (now understood to be that of closed strings) while Ziro Koba and Holar Nielsen generalized Veneziano's integral representation to be multi-particle scattering.

String Theory

Veneziano and Sergio Fubini introduced a formal operation for computing the scattering amplitudes that was a forerunner of the World-Sheet Conformal Theory, while Virasoro understood how to remove the poles with wrong-sign residues using a constraint on the states. Claud Lovelace calculated a loop amplitude, and noted that there is an inconsistency unless the dimension of the theory is 26. Charles Thorn, Peter Goddard and Richard Brower went on to prove that there are no wrong-sign propagating states in dimensions less than or equal to 26.

In 1969, Yoichiro Nambu, Holger Bech Nielsen, and Leonard Susskind recognized that the theory could be given a description in space and time in terms of strings. The scattering amplitudes were derived systematically from the action principle by Peter Goddard, Jeffrey Goldstone, Claudio Rebbi, and Charles Thorn, giving a space-time picture to the vertex operators introduced by Veneziano and Fubini and a geometrical interpretation to the Virasoro conditions.

In 1970, Pierre Ramond added fermions to the model, which led him to formulate a two-dimensional super-symmetry to cancel the wrong-sign states. John Schwarz and André Neveu added another sector to the Fermi Theory a short time later. In the fermion theories, the critical dimension was 10. Stanley Mandelstam formulated a world sheet conformal theory for both the bosonic and fermi case, giving a two-dimensional field a theoretical path to integrate a formal general operation.

String Theory

Michio Kaku and Keiji gave a different formulation of the bosonic string, as a string field theory, with infinitely many particle types and with fields taking values based not on points, but on loops and curves.

In 1974, Tamiaki Yoneya discovered that all the known string theories included a massless spin-two particle that obeyed the correct Ward Identities to be a graviton. John Schwarz and Joel Scherk came to the same conclusion and made the bold leap to suggest that string theory was a theory of gravity, not a theory of hadrons. They reintroduced the Kaluza-Klein Theory as a way of making sense of the extra dimensions. At the same time, quantum chromodynamics was recognized as the correct theory of hadrons, shifting the attention of physicists and apparently leaving the bootstrap program in the wastebasket.

String theory eventually made it out of the trash, but for the following decade all work on the theory was almost completely ignored. Still, the theory continued to develop thanks to the work of a handful of devotees. Ferdinando Gliozzi, Joel Scherk, and David Olive realized in 1976 that the original Ramond and Neveu Schwarz-strings were separately inconsistent and needed to be combined. The resulting theory did not have a tachyon, and was proven to have space-time supersymmetry by John Schwarz and Michael Green in 1981. The same year, Alexander Polyakov gave the theory a modern path integral formulation, and went on to develop conformal field theory extensively. In 1979, Daniel Friedan showed that the equations of motions of

String Theory

string theory, which are generalizations of the Einstein equations of General Relativity, emerged from the Renormalization Group equations for the two-dimensional field theory. Schwarz and Green discovered T-duality, and constructed two superstring theories -- IIA and IIB related by T-duality, and type I theories with open strings. The consistency conditions had been so strong, that the entire theory was nearly uniquely determined, with only a few discrete items left to be determined.

First Superstring Revolution

(Above) Edward Witten

In the early 1980s, Edward Witten discovered that most theories of quantum gravity could not accommodate chiral fermions like the neutrino. (Chiral means the direction of spin, see page 20.) This led him, in collaboration with Luis Álvarez-Gaumé to study violations of the conservation laws in gravity theories with anomalies, and to conclude that type I string theories were inconsistent. Green and Schwarz discovered a contribution to the anomaly that Witten and

String Theory

Alvarez-Gaumé had missed, which restricted the gauge group of the type I string theory to be SO(32). In coming to understand this calculation, Edward Witten became convinced that string theory was truly a consistent theory of gravity, and he became a high-profile advocate. Following Witten's lead, between 1984 and 1986, hundreds of physicists started to work in this field, and this is sometimes called the First Superstring Revolution.

During this period, David Gross, Jeffrey Harvey, Emil Martinec, and Ryan Rohm discovered heterotic strings. The gauge group of these closed strings was two copies of E8, and either copy could easily and naturally include the standard model. Philip Candelas, Gary Horowitz, Andrew Strominger and Edward Witten found that the Calabi–Yau manifolds (see pp. 12, 14, 41, 42) are the compactifications that preserve a realistic amount of supersymmetry, while Lance Dixon and others worked out the physical properties of orbifolds, distinctive geometrical singularities allowed in string theory. Cumrun Vafa generalized T-duality from circles to arbitrary manifolds, creating the mathematical field of mirror symmetry. Daniel Friedan, Emil Martinec and Stephen Shenker further developed the covariant quantization of the superstring using conformal field theory techniques. David Gross and Vipul Periwal discovered that string perturbation theory was divergent. Stephen Shenker showed it diverged much faster than in field theory suggesting that new non-perturbative objects were missing.

String Theory

(Above) Joseph Polchinski

In the 1990s, Joseph Polchinski discovered that the string theory required higher-dimensional objects, called D-branes (see p.17,18,27,59) and identified these with the black hole solutions of supergravity. These were understood to be the new objects suggested by the Perturbative Divergences, and they opened up a new field with rich mathematical structure. It quickly became clear that D-branes and other p-branes, not just strings, formed the matter content of the string theories, and the physical interpretation of the strings and branes revealed they were part of black holes. Leonard Susskind had incorporated the holographic principle of Gerardus 't Hooft into string theory, connecting the long, highly excited, string states with ordinary thermal black hole states. As suggested by 't Hooft, the fluctuations of the black hole horizon, the World-Sheet Theory, described not only the black hole, but all nearby objects too.

String Theory

Second Superstring Revolution

In 1995, at the annual conference of string theorists at the University of Southern California (USC), Edward Witten gave a speech on string theory that in essence united the five string theories that existed at the time, and giving birth to a new 11-dimensional theory called the M-theory. M-theory was also foreshadowed in the work of Paul Townsend at approximately the same time. The flurry of activity that began at this time is sometimes called the Second Superstring Revolution (See page 19).

(Above) Juan Maldacena

During this period, Tom Banks, Willy Fischler, Stephen Shenker, and Leonard Susskind formulated matrix theory, a full holographic description of M-theory using IIA D0 branes. This was the first definition of string theory that was fully non-perturbative and a concrete mathematical realization of the holographic principle. It is an example of

String Theory

a gauge-gravity duality and is now understood to be a special case of the AdS/CFT correspondence (See pg. 28 ff.). Andrew Strominger and Cumrun Vafa calculated the entropy of certain configurations of D-branes and found agreement with the semi-classical answer for extreme charged black holes. Petr Hořava and Witten found the eleven-dimensional formulation of the heterotic string theories, showing that orbifolds solve the chirality problem. Witten noted that the effective description of the physics of D-branes at low energies is by a supersymmetric gauge theory, and found geometrical interpretations of mathematical structures in gauge theory that he and Nathan Seiberg had earlier discovered in terms of the location of the branes.

In 1997, Juan Maldacena noted that the low energy excitations of a theory near a black hole consisted of objects close to the horizon, which for extreme charged black holes looks like an anti-de Sitter space. He noted that in this limit the gauge theory describes the string excitations near the branes. So he hypothesized that string theory on a near-horizon extreme-charged black-hole geometry, an anti-de Sitter space times a sphere with flux, is equally well described by the low-energy limiting gauge theory, the $N = 4$ supersymmetric Yang-Mills theory This hypothesis, which is called the AdS/CFT correspondence (See pg. 28 ff.) was further developed by Steven Gubser, Igor Klebanov, Alexander Polyakov, and Edward Witten, and it is now well-accepted. It is a concrete realization of the holographic principle, which has far-reaching implications for black

String Theory

holes, locality and information in physics, as well as the nature of the gravitational interaction. Through this relationship, string theory has been shown to be related to gauge theories like quantum chromodynamics (See pg. 34) and this has led to more quantitative understanding of the behavior of hadrons, bringing string theory back to its roots.

Number of Solutions

To construct models of particle physics based on string theory, physicists typically begin by specifying a shape for the extra dimensions of space time. Each of these different shapes corresponds to a different possible universe, or "vacuum state", with a different collection of particles and forces. String theory as it is currently understood has an enormous number of vacuum states, typically estimated to be around 10^{500}, and these might be sufficiently diverse to accommodate almost any phenomena that might be observed at low energies.

Many critics of string theory have expressed concerns about the large number of possible universes described by string theory. In his book *Not Even Wrong*, Peter Woit, a lecturer in the mathematics department at Columbia University, has argued that the large number of different physical scenarios renders string theory vacuous as a framework for constructing models of particle physics. According to Woit, the possible existence of, say, 10^{500} consistent different vacuum states for superstring theory probably destroys the hope of using the theory to predict anything. If one picks

String Theory

among this large set just those states whose properties agree with present experimental observations, it is likely there still will be such a large number of these that one can get just about whatever value one wants for the results of any new observation.

Some physicists believe this large number of solutions is actually a virtue because it may allow a natural anthropic explanation of the observed values of physical constants, in particular the small value of the cosmological constant. The anthropic principle is the idea that some of the numbers appearing in the laws of physics are not fixed by any fundamental principle but must be compatible with the evolution of intelligent life. In 1987, Steven Weinberg published an article in which he argued that the cosmological constant could not have been too large, or else galaxies and intelligent life would not have been able to develop. Weinberg suggested that there might be a huge number of possible consistent universes, each with a different value of the cosmological constant, and observations indicate a small value of the cosmological constant only because humans happen to live in a universe that has allowed intelligent life, and hence observers, to exist.

String theorist Leonard Susskind has argued that string theory provides a natural anthropic explanation of the small value of the cosmological constant. According to Susskind, the different vacuum states of string theory might be realized as different universes within a larger multiverse. The fact that the observed universe has a small cosmological constant is just a tautological consequence of the fact that a small value is required for life to exist. Many prominent theorists and critics have disagreed with Susskind's conclusions. According to Woit, "in this case [anthropic reasoning] is nothing more than an excuse for failure. Speculative scientific ideas fail not just when they make incorrect predictions, but also when they turn out to be vacuous and incapable of predicting anything."

Background Independence

One of the fundamental properties of Einstein's general theory of relativity is that it is background independent, meaning that the formulation of the theory does not in any way privilege a particular space time geometry.

One of the main criticisms of string theory from early on is that it is not manifestly background independent. In string theory, one must typically specify a fixed reference geometry for space time, and all other possible geometries are described as perturbations of this fixed one. In his book *The Trouble With Physics*, physicist Lee Smolin of the Perimeter Institute for Theoretical Physics claims that this is the principal weakness of string theory as a theory of quantum gravity, saying that string theory has failed to incorporate this important insight from general relativity.

Others have disagreed with Smolin's characterization of string theory. In a review of Smolin's book, string theorist Joseph Polchinski wrote:

String Theory

[Smolin] is mistaking an aspect of the mathematical language being used for one of the physics being described. New physical theories are often discovered using a mathematical language that is not the most suitable for them...In string theory it has always been clear that the physics is background-independent even if the language being used is not, and the search for more suitable language continues. Indeed, as Smolin belatedly notes, AdS/CFT (See page 28 ff.) provides a solution to this problem, one that is unexpected and powerful.

Polchinski notes that an important open problem in quantum gravity is to develop holographic descriptions of gravity which do not require the gravitational field to be asymptotically anti-de Sitter.

Smolin responded that the claims about background-independence, which Polchinski presents as "clear", are in fact only an unproven hope for future results, and Smolin is skeptical about them being true at all because of fundamental reasons, "If the strong form of the AdS/CFT conjecture is shown to be correct, then a very weak, and limited form of background will have been achieved. But this is still a big if". Smolin pointed out that current results about the AdS/CFT conjecture rely on global super-symmetry as perturbative physics, "but the whole point of general relativity and quantum gravity is that the generic solutions are governed by no global symmetries because the geometry of space time is completely dynamic," which "makes it very non-trivial to show the strong form of the AdS/CFT conjecture, because it must extend to solutions of supergravity arbitrarily far from those with global symmetries in the bulk."

Smolin summarized, "It would be more accurate to say, 'Some string theorists believe that the formulations of perturbative string theories and dualities between them that they study concretely are approximations to a deeper, background independent formulation. This missing background independent formulation is not just a different language for the theory, it is hoped to be the statement of the principles and laws that define the theory, from which everything studied so far would be derived as an approximation.'"

Sociological Issues

Since the superstring revolutions of the 1980s and 1990s, string theory has become the dominant paradigm of high energy theoretical physics. Some string theorists have expressed the view that there does not exist an equally successful alternative theory addressing the deep questions of fundamental physics. In an interview from 1987, Nobel Laureate David Gross made the following controversial comments about the reasons for the popularity of string theory:

"The most important [reason] is that there are no other good ideas around. That's what gets most people into it. When people started to get interested in string theory they didn't

String Theory

know anything about it. In fact, the first reaction of most people is that the theory is extremely ugly and unpleasant, at least that was the case a few years ago when the understanding of string theory was much less developed. It was difficult for people to learn about it and to be turned on. So I think the real reason why people have got attracted by it is because there is *no other game in town*. All other approaches of constructing grand unified theories, which were more conservative to begin with, and only gradually became more and more radical, have failed, and this game hasn't failed yet."

Several other high profile theorists and commentators have expressed similar views, suggesting that there are no viable alternatives to string theory.

Many critics of string theory have commented on this state of affairs. In his book criticizing string theory, *Not Even Wrong,* Peter Woit viewed the status of string theory research as unhealthy and detrimental to the future of fundamental physics. He argued that the extreme popularity of string theory among theoretical physicists is partly a consequence of the financial structure of academia and the fierce competition for scarce resources.

In his book *The Road to Reality*, mathematical physicist Roger Penrose expressed similar views, stating, "The often frantic competitiveness that this ease of communication engenders leads to *bandwagon* effects, where researchers fear to be left behind if they do not join in." Penrose also

String Theory

claimed that the technical difficulty of modern physics forces young scientists to rely on the preferences of established researchers, rather than forging new paths of their own.

Lee Smolin expressed a slightly different position in his critique, claiming that string theory grew out of a tradition of particle physics which discouraged speculation about the foundations of physics, while his preferred approach, loop quantum gravity, encourages more radical thinking.

According to Smolin, "String theory is a powerful, well-motivated idea and deserves much of the work that has been devoted to it. If it has so far failed, the principal reason is that its intrinsic flaws are closely tied to its strengths -- and, of course, the story is unfinished, since string theory may well turn out to be part of the truth. The real question is not why we have expended so much energy on string theory but why we haven't expended nearly enough on alternative approaches." Smolin went on to offer a number of prescriptions for how scientists might encourage a greater diversity of approaches to quantum gravity research.

String Theory

Chapter 6. Beyond Nanotech to Femtotech

But nano is not as small as the world can go. A nanometer is 10^{-9} meters – the scale of atoms and molecules. A water molecule is a bit less than one nanometer long, and a germ is around a thousand nanometers across. On the other hand, a proton has a diameter of a couple of femtometers – where a femtometer, at 10^{-15} meters, makes a nanometer seem positively gargantuan. Now that the viability of nanotech is widely accepted (in spite of some ongoing heated debates about the details), it's time to ask: what about femtotech? Picotech or other technologies at the scales between nano and femto seem relatively uninteresting, because we don't know any basic constituents of matter that exist at those scales. But femtotech, based on engineering structures from subatomic particles, makes perfect conceptual sense, though it's certainly difficult given current technology.

The nanotech field was arguably launched by Richard Feynman's 1959 talk "There's Plenty of Room at the Bottom." As Feynman wrote (See chapter 1),

String Theory

"It is a staggeringly small world that is below. In the year 2000, when they look back at this age, they will wonder why it was not until the year 1960 that anybody began seriously to move in this direction".

But Feynman's original vision, while it was focused on the nano-scale, wasn't restricted to this level. "There's plenty of room at the bottom," he said, "and the nano-scale is not the bottom! There's plenty more room down there to explore."

One might argue that, since practical nanotech is still at such an early stage, it's not quite the time to be thinking about femtotech." But technology is advancing faster and faster each year, so it makes sense to think a bit further ahead than contemporary hands-on engineering efforts." Various scientific books and journals say that the topic is worth looking at in spite of our current lack of knowledge regarding its practical realization. After all, when Feynman gave his "Plenty of Room at the Bottom" lecture, nanotech also appeared radically pie-in-the-sky.

There are many possible routes to femtotech. Here we'll focus on a particular class of approaches to femtotech based on the engineering of stable degenerate matter – not because this is the only interesting way to think about femtotech, but merely because one has to choose some definite direction to explore if one wants to go into any detail at all.

String Theory

Physics at the Femto Scale

To understand the issues involved in creating femtotech, you'll need to learn a few basics about particle physics.

In the picture painted by contemporary physics, everyday objects like houses and people and water are made of molecules, which are made of atoms, which in turn are made of subatomic particles. There are also various subatomic particles that don't form parts of atoms (such as photons, the particles of light, and many others). The behavior of these particles is extremely weird by the standards of everyday life – with phenomena like non-local correlations between distant phenomena, observer-dependence on reality, quantum teleportation and lots of other things. But here a few facts about subatomic particles will be discussed which lead to an explanation of how femtotech might come about.

Subatomic particles fall into two categories: fermions and bosons. These two categories each contain pretty diverse sets of particles, but they're grouped together because they also have some important commonalities.

The particles that serve as the building blocks of matter are all fermions. Atoms are made of protons, neutrons and electrons. Electrons are fermions, and so are quarks, which combine to build protons and neutrons. Quarks appear to occur in nature only in groups, most commonly in groups of 2 or 3. A proton contains two up quarks and one down quark, while a neutron consists of one up quark and two down

String Theory

quarks; the quarks are held together in the nucleus by other particles called gluons. Mesons consist of two quarks – a quark and an anti-quark. There are six basic types of quarks, named Up, Down, Bottom, Top, Strange, and Charm. Out of the four forces currently recognized in the universe – electromagnetism, gravity and weak and strong nuclear forces – quarks are most closely associated with the strong nuclear force, which controls most of their dynamics. But quarks also have some interaction with the weak force, e.g. the weak force can cause the transmutation of quarks into different quarks, a phenomenon that underlies some kinds of radioactive decay such as beta decay.

On the other hand, bosons are also important – for example photons, the particle-physics version of light, are bosons. Gravitons, the gravity particles proposed by certain theories of gravitation, would also be bosons.

The nucleus of an atom contains protons and neutrons. The electrons are arranged in multiple shells around the nucleus, due to the Pauli exclusion principle. Also note this sort of "solar system" model of particles as objects orbiting other objects is just a heuristic approximation; there are many other complexities and a more accurate view would depict each particle as a special sort of wave function.

There are also carbon atoms, whose electrons are distributed across two shells.

Finally, fermions, unlike bosons, obey the Pauli exclusion principle, which says that no two identical fermions can occupy the same state at the same time. For example, each electron in an atom is characterized by a unique set of quantum numbers (the principle quantum number which gives its energy level, the magnetic quantum number which gives the direction of orbital angular momentum, and the spin quantum number which gives the direction of its spin). If not for the Pauli exclusion principle, all of the electrons in an atom would pile up in the lowest energy state (the K shell, the innermost shell of electrons orbiting the nucleus of the atom). But the exclusion principle implies that the different electrons must have different quantum states, which results in some of the electrons getting forced to have different positions, leading to the formation of additional shells (in atoms with sufficient electrons).

Degenerate Matter as a Possible Substrate for Femtotech

One can view the Pauli exclusion principle as exerting a sort of "pressure" on matter, which in some cases serves to push particles apart. In ordinary matter this Pauli pressure is minimal compared to other forces. But there is also degenerate matter – matter which is so extremely dense that this Pauli pressure or "degeneracy pressure", preventing the constituent particles from occupying identical quantum states, plays a major role. In this situation, pushing two particles close together means that they have effectively identical positions, which means that in order to obey the Pauli exclusion principle, they need to have different energy

String Theory

levels, creating a lot of additional compression force, and causing some very odd states of matter to arise.

For instance, in ordinary matter, temperature is correlated to speed of molecular motion. Heat implies faster motion, and cooling something down makes its component molecules move more slowly. But in degenerate matter, this need not be the case. If one repeatedly cools and compresses a plasma, eventually one reaches a state where it's not possible to compress the plasma any further, because of the exclusion principle that won't let us put two particles in the same state (including the same place). In this kind of super-compressed plasma, the position of a particle is rather precisely defined – but according to a key principle of quantum theory, Heisenberg's Uncertainty Principle, you can't have accurate knowledge of both the position and the momentum (movement) of a particle at the same time. So the particles in a super-compressed plasma must therefore have highly uncertain momentum – i.e. in effect, they're moving around a lot, even though they may still be very cold. This is just one example of how degenerate matter can violate our usual understanding of how materials work.

At the present time, degenerate matter is mostly discussed in astrophysics, in the context of neutron stars, white dwarf stars, and so forth. It has also been popular in science fiction – for example, in the Star Trek universe, neutronium (matter formed only from large numbers of neutrons, stable at ordinary gravities) is an extremely hard and durable substance, often used as armor, which conventional weapons cannot penetrate or dent at all. But so far neutronium has never been seen in reality. "Strange matter" – defined as matter consisting of an equal number of up, down and strange quarks – is another kind of degenerate matter, with potential applications to femtotech, which will be discussed a little later.

As a substrate for femtotech, degenerate matter appears to have profound potential. It serves as an existence proof that, yes, one can build stuff other than atoms and molecules with subatomic particles. On the other hand, there is the problematic fact that all the currently known examples of degenerate matter exist at extremely high gravities, and derive their stability from this extreme gravitational force. Nobody knows, right now, how to make degenerate matter that remains stable at Earth-level gravities or anywhere near. However, neither has anybody shown that this type of degenerate matter is an impossibility according to our currently assumed physical laws. It remains a very interesting open question.

What Will Femtotechnology Computers Look-Like?

How the properties of quarks and gluons can be used (in principle) to perform computations at the femtometer (10^{-15} meter) scale. 29 October 2011, by Hugo de Garis

I've been thinking on and off for two decades about the possibility of a femtotech. Now that nanotech is well

String Theory

established, and well funded, I feel that the time is right to start thinking about the possibility of a femtotech.

You may ask, "What about picotech?" -- technology at the picometer (10^{-12}m) scale. The simple answer to this question is that nature provides nothing at the picometer scale. An atom is about 10^{-10} meter in size.

The next smallest thing in nature is the nucleus, which is about 100,000 times smaller, i.e., 10^{-15} m in size -- a femtometer, or "fermi." A nucleus is composed of protons and neutrons (i.e., "nucleons"), which we now know are composed of 3 quarks, which are bound ("glued") together by massless (photon-like) particles called "gluons."

Hence if one wanted to start thinking about a possible femtotech, one would probably need to start looking at how quarks and gluons behave, and see if these behaviors might be manipulated in such a way as to create a technology, i.e., computation and engineering (building things).

The author of this essay, Hugo de Garis, concentrated on the computation side, since his background was in computer science. Before he retired, he was a computer science professor who gave himself zero chance of getting a grant from conservative NSF or military funders in the U.S. to speculate on the possibilities of a femtotech. But when he was no longer a wage earner, he was free to do what he liked and could join the billion strong "army" of post retirement people to pursue his own passions.

String Theory

So he started studying QCD (quantum chromodynamics), the mathematical physics theory of the strong force, or as it is known in more modern terms, the "color force."

Since he had a computer science background, he knew what to look for when sniffing through QCD text books, to be able to map computer science concepts to QCD phenomena.

Bits & Logic Gates: The Heart of Computation

If you want to compute at the femto level, how do you do that? What would you need? To de Garis, the essential ingredients of (digital) computing were bits and logic gates.

A bit is a two-state system (e.g., voltage or no voltage, a closed or open switch, etc.) that could be switched from one state to another. It is usual to represent one of these states as "1" and the other as "0," i.e., as binary digits. A logic gate is a device that can take bits as input and use their states (their 0 or 1 values) to calculate its output.

The three most famous gates, are the NOT gate, the OR gate, and the AND gate. The NOT gate switches a 1 to a 0, and a 0 to a 1. An OR gate outputs a 1 if one or more of its two inputs is a 1, else outputs a 0. An AND gate outputs a 1 only if the first AND second inputs are both 1, else outputs a 0.

There is a famous theorem in theoretical computer science, that says that the set of 3 logic gates {NOT, OR, AND} are "computationally universal," i.e., using them, you can build

String Theory

any Boolean logic gate to detect any Boolean expression (e.g. (~X & Y) OR (W & Z)).

So de Garis said, "If I can find a one to one sitting between these 3 logic gates and phenomena in QCD, I can compute anything in QCD. I would have femtometer-scale computation." So, he set out to find phenomena in QCD that he could map bits and logic gates to. He was quickly rewarded.

The Color Charge On Quarks and Gluons

There are four types of force in the physical world, from weakest to strongest: the gravitational force, the weak nuclear force, the electromagnetic force, and the strong nuclear force. (Really, their relative strengths depend on the temperature at which these forces act. At the extreme temperatures (energies) that occurred just after the big bang and now at the LHC (Large Hadron Collider) in Geneva, their strengths converge to the same value, a phenomenon called "grand unification."

In the 60s and 70s physicists became aware that the nucleons (the protons and the neutrons) consisted of 3 quarks, which had fractional electric charges (e.g., +/- 1/3 or 2/3 of the charge of an electron), and a new type of charge, called "color." The electronic charge came in two types (positive and negative), which is something science has known about for several centuries. The color charge however comes in 3 types, "red" "blue" and "green."

String Theory

The electromagnetic force is "mediated" (conveyed) between two electrical charges via the photon (the particle of light). A photon is emitted by one of the charges and is absorbed by the other. This interaction creates the attractive or repulsive force between the electrical charges.

Something similar happens between quarks. The equivalent of the photon is called a gluon. A quark emits a gluon, which is then absorbed by another quark, and this creates the interaction between the two quarks.

There is an essential difference between a photon and a gluon. The photon has no charge of its own, whereas a gluon does have a color charge; in fact, each gluon has two such charges. It is bi-charged, or bi-colored. This means that gluons can interact with other gluons, forming complex "glueballs." "Glueballs" were not be used in this article, but they might play an important role in femtotech in the future.

Strictly speaking, there are 6 color charges, namely red, blue, green, anti-red, anti-blue, and anti-green. A gluon (at least the type of gluon used in this essay) has one of the first three color charges, and one of the second three. So there are 6 such bi-colored gluons, a red, anti-blue; a red, anti-green; a blue, anti-red; a blue, anti-green; a green, anti-red; and a green, anti-blue. In this essay only the red, anti-blue and the blue, anti-red gluons will be used, because, using Occam's razor, they are all that I need. Occam's razor (or Ockham's) is a line of reasoning that says the simplest answer is often correct, according to *howstuffworks.com* on the internet.

String Theory

Quark-Gluon Reactions Conserve Colors

How does a gluon interact with a quark? What happens? Remarkably, when a gluon and a quark interact, the gluon may change the quark's color, and in such a way that the colors are conserved. For example, imagine a red, anti-blue gluon (which from now on will be abbreviated to Gr,~b) interacts with a blue-colored quark (abbreviated from now on to Qb). The gluon will cause the quark to change its color from blue to red, i.e., in symbolic terms:

Gr,~b : Qb -> Qr

In other words, the red, anti-blue gluon acts on the blue (color charged) quark, and converts it into a red (color charged) quark.

Note that before the interaction, there were 3 charges: a red, an anti-blue (both on the gluon), and a blue (on the quark). During the interaction, the anti-blue of the gluon and the blue of the quark cancel, leaving only a red, which is now the color (charge) of the outgoing quark. The colors are conserved.

What would happen if a red, anti-blue gluon (Gr,~b) interacted with a red quark Qr? Nothing. Such an interaction is forbidden in nature, because the color charges in this case are not conserved. Before the interaction, we have a red and an anti-blue charge on the gluon, and a red on the quark. If the quark absorbed the gluon and changed its color from red to blue, then the final charge would be just

String Theory

blue. But that doesn't match the "2 reds and 1 anti-blue charges" before the interaction. The colors are not conserved, so this interaction is QCD forbidden.

This color conservation operates with the emission of a gluon as well. For example, a red quark Qr could emit a red, anti-blue gluon (Gr,~b) and become a blue quark (Qb). This emission can be represented as

Qr -> Qb + Gr,~b

Note that the colors are conserved. The blue and anti-blue cancel each other, leaving a red on both sides. Color conservation is one of the basic natural laws of QCD.

Now, a gluon that is emitted by one quark can be absorbed by another quark, rather like the way a photon can be emitted and absorbed by two electrically charged particles (which is the basis of the study of quantum electrodynamics, QED). By emitting and absorbing gluons, two quarks can interact with each other and influence each other.

The "aha moment"

Probably some of you have already had an "aha moment" on how you might implement femtotech-based computing, based only on what was said above.

Hugo de Garis, the author of these pages, said he had an "aha moment" when he realized the color conservation rules given above. He felt he had found a way to compute at the

String Theory

femtometer scale, using quarks and gluons, at least in principle. For difficulties facing the practical engineering of these ideas, see towards the end of this essay.

The aha moment gave him the following basic ideas.

- Represent a bit by the color of a quark. A red for 1, and a blue for 0. (He didn't need to use green.)
- To change the state (1 to 0, or 0 to 1), change the color of the quark from red to blue, or vice versa.
- To change the color of a quark, use an appropriately emitted gluon, i.e., one possessing the appropriate bi-coloring.
- To implement logic gates (and this was the creative challenging part), use a sequence of gluon emission and absorption (of the same gluon).

Mapping the Gates to Quark-Gluon Interactions

Next he needed to introduce a fictional didactic device that he called a "quark chamber," i.e., a region of space (perhaps as small as a nucleon), such as a sphere, in which a quark enters at one end, interacts (or fails to interact), and exits at the other end. Also entering or exiting the quark chamber is a gluon. In the case of gluon emission, the gluon exits the quark chamber. In the case of gluon absorption, the gluon enters the quark chamber and is absorbed within it.

String Theory

The NOT Gate

Fill the quark chamber with two gluons: a Gr,~b and a Gb,~r. If a red quark Qr enters the quark chamber, it will not interact with the Gr,~b gluon, but will be converted to a blue quark by absorption of a Gb,~r gluon, and will exit the quark chamber as a blue quark, according to the interaction below:

Gb,~r : Qr -> Qb

An ipso facto interaction will occur for a blue quark entering the quark chamber, according to the interaction below:

Gr,~b : Qb -> Qr

We thus have a NOT gate. A red quark is converted to a blue quark (1 to 0), and a blue quark is converted to a red quark (0 to 1). This is the definition of a NOT gate.

The OR Gate

To implement an OR gate is a bit more complicated. We need 2 quark chambers, A, B. Chamber A is a gluon generating chamber. If a red quark enters chamber A, a red, anti-blue gluon Gr,~b emission is caused in the chamber and the gluon then exits. (The resulting blue quark is ignored.)

If a blue quark enters chamber A, nothing happens. No gluon exits the chamber.

String Theory

We now have 4 cases to consider:

a) red(1), red(2): (a red quark(1) enters chamber A, and a second red quark(2) enters chamber B). The red quark Qr(1) entering chamber A generates a Gr,~b gluon that enters chamber B. This gluon has no effect on the red Qr(2) entering chamber B at the same time. The red Qr(2) then passes out of chamber B unaffected. In other words, the output quark from chamber B is red. Hence if the inputs are red(1) and red(2) the output quark is red.

b) red(1), blue(2): The red quark Qr(1) entering chamber A generates a Gr,~b gluon that enters chamber B. The blue quark Qb(2) that enters chamber B is converted to a red quark Qr(2) that then exits chamber B. In other words, the output quark from chamber B is red. Hence if the inputs are red(1) and blue(2) the output quark is red.

c) blue(1), red(2): The blue quark Qb(1) entering chamber A generates NO gluon, so no gluon enters chamber B. The red quark Qr(2) that enters chamber B then exits unchanged. In other words, the output quark from chamber B is red. Hence if the inputs are blue(1) and red(2) the output quark is red.

d) blue(1), blue(2): The blue quark Qb(1) entering chamber A generates NO gluon, so no gluon enters chamber B. The blue quark Qb(2) that enters chamber B then exits chamber B unchanged. In other words, the output quark from chamber B is blue. Hence if the inputs are blue(1) and blue(2) the output quark is blue.

String Theory

Thus the specifications of an OR gate are satisfied.

The AND Gate

The AND gate is a bit more complicated. It contains 3 chambers, A, B, C. Chambers A and B both output a red quark if the input is a red quark, and a blue, anti-red gluon Gb,~r if the input is a blue quark. This time, instead of dealing with single events, think in terms of a stream of input and output quarks. Chamber C has as input, the outputs of chambers A and B, as well as a fixed red quark Qr(3) input, for reasons that will soon become clear.

We again have 4 cases to consider:

a) red(1), red(2): (red quarks(1) enter chamber A, and red quarks(2) enter chamber B). The red quarks Qr(1) and Qr(2) pass unchanged into chamber C, along with the fixed red quarks Qr(3). There are only red quarks in chamber C, so only red quarks can exit chamber C. In other words, the output quarks from chamber C are red. Hence if the inputs are red(1) and red(2) the output quarks are red (now thinking in terms of streams of quarks).

b) red(1), blue(2): (red quarks(1) enter chamber A, and blue quarks(2) enter chamber B). The red quarks Qr(1) pass unchanged into chamber C, along with the fixed red quarks Qr(3). The blue quarks Qb(2) that enter chamber B generate blue, anti-red gluons Gb,~r which pass into chamber C. These gluons convert all the red quarks in chamber C to blue

String Theory

quarks, so that only blue quarks exit from chamber C. Hence if the inputs are red(1) and blue(2) the output quarks are blue.

c) blue(1), red(2): (blue quarks(1) enter chamber A, and red quarks(2) enter chamber B). The blue quarks Qb(1) that enter chamber A generate blue, anti-red gluons Gb,~r which pass into chamber C. The red quarks Qr(2) that enter chamber B pass unchanged into chamber C, along with the fixed red quarks Qr(3). These gluons convert all the red quarks in chamber C to blue quarks, so that only blue quarks exit from chamber C. Hence if the inputs are blue(1) and red(2) the output quarks are blue.

d) blue(1), blue(2): (blue quarks(1) enter chamber A, and blue quarks(2) enter chamber B). The blue quarks Qb(1) and Qb(2) both generate blue, anti-red gluons Gb,~r which pass into chamber C. These gluons convert the fixed red quarks entering chamber C to blue quarks, so that only blue quarks exit from chamber C. Hence if the inputs are blue(1) and blue(2) the output quarks are blue.

Thus the specifications of an AND gate are satisfied.

Engineering Challenges

Now that all 3 gates have been mapped to quark-gluon interactions in QCD, one has an "in principle" recipe for femtometer scale computation.

However the practical engineering problems remain, especially when considering something called "asymptotic

String Theory

freedom," which says that quarks interact weakly when close together, but immensely strongly as they separate, rather like a tough rubber band being stretched. The more it is stretched, the greater the potential energy it has. Similarly with the 3 quarks inside a nucleon.

A nucleon is stable (in the nucleus) because it has 3 quarks, one is red, another blue, and the third green. These 3 colors "sum" to "white" (rather like a spinning color wheel of equally sized red, blue and green sectors), which is analogous to the way an atom, with its positively charged nucleus and its negatively charged electrons, sums to neutrality.

However, if one attempts to extract a quark from the nucleon, the gluons between the extracting quark and the other two quarks, behave in complex nonlinear ways, interacting with other gluons, to form a hugely powerful resistance, until the potential energy is so great that a quark, anti-quark pair can be formed, which combine to form a pion (pi meson). (Mesons consist of 2 quarks: a quark and its anti-quark.) Hence it seems impossible to isolate a quark (or a gluon). Experimentally, no quark or gluon has ever been isolated. Experimentalists have virtually given up trying.

Hence the implicit assumption in the above models, namely that isolated quarks and gluons are used, seems unphysical and unrealistic.

But, if the gluons and quarks are close together, the "stretching rubber band" phenomenon does not occur. There

String Theory

may be particles that contain more than 3 quarks, the so called "exotics," which may have 3N quarks (a multiple of 3 to maintain color neutrality ("whiteness") by summing an equal number of red, blue, and green color charges).

There may also be "glueballs" that consist only of gluons that interact in highly non-linear and hence complex ways.

Another possibility is to heat up the quark/gluon complex so much that one obtains a quark-gluon "plasma." At a critical temperature, after cooling the plasma, quark-gluon "chains" may start forming, that may interact in ways similar to the way molecules interact within the cell, i.e., by complementary "lock and key" touching.

Conclusions of Hugo de Garis

The above femtometer scale computation models are "in principle" only. To make them practical will probably require new thinking, to ensure that they are compatible with the severe constraints imposed by the principles of QCD, e.g., quark confinement and asymptotic freedom.

Hopefully, this essay will stimulate other researchers to enter this new research field of femtotech. Perhaps the "other side" of technology (the "building stuff" side, in contrast with the computational side) can be implemented with glueballs as well, or with quark/gluon "condensates."

One thing is clear. If humanity does not make any progress along the lines of femtotech, sooner or later, human beings

String Theory

(or our artificially intelligent successors) will be scratching at the "nanotech walls" that confine us.

As one final comment, Hugo de Garis, the author of these pages, was thinking of trying to create an "attotech" (i.e., on the scale of 10^{-18} meters) by using the weak-force particles (W and Z particles) that interact not only with quarks, but with the much lighter leptons (e.g., electrons, etc.) as well.

Human technology has progressed from millitech, to microtech, to (recently) nanotech, and this essay attempts to start the thinking on femtotech (and attotech).

This downscaling trend provides a potential answer to the famous "Fermi paradox" (if intelligent life is so commonplace in the universe, "where are they?"). If intelligent creatures or machines can continue to "scale down" in their technologies, the answer to Fermi's question would become "They are all around us, whole civilizations living inside elementary particles, too small for us to detect."

The reference for the above pages can be found in Appendix A, at the very end. Reference: 141.

String Theory

String Theory

Chapter 7. Femtotechnology 2016

Femtotechnology is a hypothetical term used in reference to structuring of matter on the scale of a femtometer, which is 10^{-15} m. This is a smaller scale in comparison to nanotechnology and picotechnology which refer to 10^{-9} m and 10^{-12} m respectively.

Theory

Work in the femtometer range involves manipulation of excited energy states within atomic nuclei, specifically nuclear isomers, to produce metastable (or otherwise stabilized) states with unusual properties. In the extreme case, excited states of the individual nucleons that make up the atomic nucleus (protons and neutrons) are considered, ostensibly to tailor the behavioral properties of these particles.

The most advanced form of molecular nanotechnology is often imagined to involve self-replicating molecular machines, and there have been some speculations suggesting something similar might be possible with analogues of molecules composed of nucleons rather than atoms. For example, the astrophysicist Frank Drake once speculated about the possibility of self-replicating organisms composed

String Theory

of such nuclear molecules living on the surface of a neutron star, a suggestion taken up in the science fiction novel *Dragon's Egg* by the physicist Robert Forward. It is thought by physicists that nuclear molecules may be possible, but they would be very short-lived, and whether they could actually be made to perform complex tasks such as self-replication, or what type of technology could be used to manipulate them, is unknown.

Practical applications of femtotechnology are currently considered to be unlikely. The spacings between nuclear energy levels require equipment capable of efficiently generating and processing gamma rays, without equipment degradation. The nature of the strong interactions is such that excited nuclear states tend to be very unstable (unlike the excited electron states in Rydberg atoms), and there are a finite number of excited states below the nuclear binding energy, unlike the (in principle) infinite number of bound states available to an atom's electrons. Similarly, what is known about the excited states of individual nucleons seems to indicate that these do not produce behavior that in any way makes nucleons easier to use or manipulate, and indicates instead that these excited states are even less stable and fewer in number than the excited states of atomic nuclei. The hypothetical hafnium bomb can be considered a crude application of femtotechnology.

(The hafnium controversy is a debate over the possibility of 'triggering' rapid energy releases, via gamma ray emission, from a nuclear isomer of hafnium. The energy release is potentially five orders of magnitude more energetic than a chemical reaction but three orders of magnitude less than a nuclear reaction. In 1998, a group at the University of Texas at Dallas reported having successfully initiated such a trigger. To date, no other group has been able to duplicate these results).

String Theory

String Theory

Chapter 8. Femtotech: (Sub)Nuclear Scale Engineering and Computation

This whole chapter is the work of an Internet writer who has some interesting and unusual ideas.

Bolonkin's Fantastic Femtotech Designs

If you type "femtotech" into a search engine, you'll likely come up with a 2009 paper by A.A. Bolonkin, a former Soviet physicist now living in Brooklyn, entitled "Femtotechnology: Nuclear Matter with Fantastic Properties." Equations and calculations notwithstanding, this is an explicitly speculative paper – but the vision it presents is intriguing.

Bolonkin describes a new (and as yet unobserved) type of matter he calls "AB-matter", defined as matter which exists at ordinary Earth-like gravities, yet whose dynamics are largely guided by the Pauli exclusion principle based degeneracy pressure. He explores the potential of creating threads, bars, rods, tubes, nets and so forth using AB-matter.

He argues that, *"this new 'AB-Matter' has extraordinary properties (for example, tensile strength, stiffness, hardness,*

String Theory

critical temperature, superconductivity, super-transparency and zero friction), which are up to millions of times better than corresponding properties of conventional molecular matter. He shows concepts of design for aircraft, ships, transportation, thermonuclear reactors, constructions and so on from nuclear matter. These vehicles will have unbelievable possibilities (e.g., invisibility, ghost-like penetration through any walls and armor, protection from nuclear bomb explosions and any radiation flux)."

Gell-Mann Gives Femtotech a Definite Maybe

The author of the above review of Bolonkin's ideas, while at an event in San Francisco, was thrilled to have the opportunity to discuss femtotech with Murray Gell-Mann who is not only a Nobel Prize winning physicist, but also one of the world's ultimate gurus on quarks, since he invented and named the concept and worked out a lot of the theory of their behavior. The reviewer's friend, Hugo de Garis, had briefly discussed femtotech with Gell-Mann a decade and a half previously, but he (Gell-Mann) hadn't expressed any particular thoughts on the topic. The reviewer was curious if Gell-Mann's views on the topic had perhaps progressed a bit.

To his mild disappointment, Gell-Mann's first statement to him about femtotech was that he had never thought about the topic seriously. However, he went on to say that it seemed to be a reasonable idea to pursue.

String Theory

When he probed Gell-Mann about degenerate matter, Gell-Mann spent a while musing about the possible varieties of degenerate matter in which the ordinary notion of quark confinement is weakened. "Confinement" is the property that says quarks cannot be isolated singularly, and therefore cannot be directly observed, but can only be observed as parts of other particles like protons and neutrons. At first it was thought that quarks could only be observed in triplets, but more recent research suggested the possibility of "weak confinement" that allows observation of various aspects of individual quarks in an isolated way. Quark-gluon plasmas, which have been created in particle accelerators using very high temperatures (like 4 trillion degrees), are one much-discussed way of producing "almost unconfined" quarks. But Gell-Mann felt the possibilities go far beyond quark-gluon plasmas. He said he thought it possible that large groups of quarks could potentially be weakly confined in more complex ways that nobody now understands.

So after some discussion in this vein, the above author pressed Gell-Mann specifically on whether understanding these alternative forms of weak multi-quark confinement might be one way to figure out how to build stable degenerate matter at Earth gravity. His answer was, basically, definitely maybe.

Eric Drexler's Ideas

The next big step toward nanotech was Eric Drexler's classic 1992 book *Nanosystems*, which laid out conceptual designs

String Theory

for a host of nanomachines, including nanocomputer switches, general-purpose molecular assemblers, and an amazing variety of other items. Drexler's 1987 book *Engines of Creation* also played a large role, bringing the notion of nanotech to the masses. Contemporary nanotech mostly focuses on narrower nano-engineering than what Drexler envisioned, but arguably it's building tools and understanding that will ultimately be useful for realizing Feynman's and Drexler's vision. For instance, a lot of work is now going into the manufacture and utilization of carbon nanotubes, which have a variety of applications, from the relatively mundane (e.g. super-strong fabrics and fibers) to potential roles as components of more transformative Nano systems like Nano computers or molecular assemblers. And there are also a few labs such as Zyvex that are currently working directly in a Drexlerian direction.

Might Dynamic Stabilization Work on Degenerate Matter?

The internet author of the paragraphs above put forward some avant guarde ideas. Here is his rather interesting brainstorm. Where he says "I" he means himself, not the author of this book.

I had a thought about how to stabilize degenerate femto-matter: use dynamic stabilization. The classic example is the shaking inverted pendulum. An upside down pendulum is unstable, falling either left or right if perturbed. But if you shake the base at a sufficiently high frequency, it adds a

String Theory

"ponder-motive" pseudopotential which stabilizes the unstable fixed point.

The same approach can stabilize fluid instabilities. If you turn a cup of fluid upside down, the perfectly flat surface is an unstable equilibrium. The Rayleigh-Taylor instability causes ripples to grow and the fluid to spill out. But, I remember seeing a guy years ago who put a cup of oil in a shaking apparatus and was able to turn it upside down without it spilling. So the oscillations were able to stabilize all the fluid modes at once. I wonder if something similar might be used to stabilize degenerate matter at the femto scale?

A fascinating idea indeed! Instead of massive gravity or massive heat, perhaps one could use incredibly fast, low-amplitude vibrations to stabilize degenerate matter. How to vibrate subatomic particles that fast is a whole other matter, and surely a difficult engineering problem – but still, this seems a quite promising avenue. It would be interesting to do some mathematics regarding the potential dynamic stabilization of various configurations of subatomic particles subjected to appropriate vibrations.

An inverted pendulum kept vertical via dynamic stabilization. The rod would rotate and fall down to one side or another if it weren't vibrating. But if it's vibrated very fast with low amplitude, it will remain upright due to dynamic stabilization. Conceivably a similar phenomenon

String Theory

could be used to make stable degenerate matter, using very fast femto-scale vibrations.

Of course, such metaphorical ideas must be taken with a grain of salt. When I think about the "liquid drop" model of the nucleus, I'm somewhat reminded of how the genius inventor Nikola Tesla intuitively modeled electricity as a fluid. This got him a long way compared to his contemporaries, leading him to develop AC power and ball lightning generators and all sorts of other amazing stuff – yet it also led to some mistakes, and caused him to miss some things that are implicit in the mathematics of electromagnetism but not in the intuitive metaphorical "electricity as fluid" model. For instance, Tesla's approach to wireless power transmission was clearly misguided in some respects (even if it did contain some insights that haven't yet been fully appreciated), and this may have been largely because of the limitations of his preferred fluid-dynamics metaphor for electricity. Where degenerate matter is concerned, metaphors to liquid drops and macroscopic shaking apparatuses may be very helpful for inspiring additional experiments, but eventually we can expect rigorous theory to far outgrow them.

The bottom line is, in the current state of physics, nobody can analytically solve the equations of nuclear physics except in special simplified cases. Physicists often rely on large-scale computer simulations to solve the equations in additional cases – but these depend on various technical simplifying assumptions, which are sometimes tuned based

String Theory

on conceptual assumptions about how the physics works. Intuitive models like "nucleus as water droplet" are based on the limited set of cases in which we've explored the solutions of the relevant equations using analytical calculations or computer simulations. So, based on the current state of the physics literature, we really don't know if it's possible to build stable structures of the sort Bolonkin envisions. But these are surely worthwhile avenues to explore.

Then we changed the topic to AI (Artificial Intelligence) and the singularity, where I'm on firmer ground – and there he was a little more positive, actually. He said he thought it was crazy to try to place a precise date on the singularity, or to estimate anything in detail about it in advance... but he was sympathetic to the notion of accelerating technological change, and very open to the idea that massively more change is on the way. And, contrary to his fellow physicist Roger Penrose, he expressed doubt that quantum computing (let alone femto-computing) would be necessary for achieving human-level AI. Even if the human brain somehow uses strange quantum effects in some particulars, he felt, digital computers should most likely be enough to achieve human-level intelligence.

A few moments later at the same event, I asked a young Caltech physics postdoc the same questions about degenerate matter and femtotech – and he gave a similar answer, only mildly more negative in tone. He said it seemed somewhat unlikely that one could make room-temperature stable structures using degenerate matter, but that he couldn't think of any strong reason why it would be impossible.

Currently, it seems, where degenerate matter based femtotech is concerned, nobody knows...

Strange Matter and Other Strange Matters

Gell-Mann's comments reminded me of strangelets – strange hypothetical constructs I first found out about a few years ago when reading about some strange people who had the strange idea that the Large Hadron Collider might destroy the world by unleashing a strange chain reaction turning the Earth into strangelets. Fortunately, this didn't happen – and it seems at least plausible that strangelets might pose a route to stable degenerate matter of a form useful for femtotech.

A strangelet is (or would be, if they exist at all, which is unknown) an entity consisting of roughly equal numbers of up, down and strange quarks. A small strangelet would be a few femtometers across, with around the mass of a light nucleus. A large strangelet could be meters across or more, and would then be called a "strange star" or a "quark star."

In a (hypothetical) strange star, quarks are not confined in the traditional sense, but may still be thought of as "weakly confined" in some sense (at least that was this Internet author's view).

String Theory

So far, all the known particles with strange quarks – like the lambda particle – are unstable. But there's no reason to believe that states with a larger number of quarks would have to suffer from this instability. According to Bodner and Witten's "strange matter hypothesis," if enough quarks are collected together, you may find that the lowest energy state of the collective is a strangelet, i.e. a state in which up, down, and strange quarks are roughly equal in number.

So where does the End of the World come in? There are some interesting (albeit somewhat speculative) arguments to the effect that if a strangelet encounters ordinary matter, it could trigger a chain reaction in which the ordinary matter gets turned into strangelets, atom by atom at an accelerating pace. Once one strangelet hits a nucleus, it would likely turn it into strange matter, thus producing a larger and more stable strangelet, which would in turn hit another nucleus, etc. Goodbye, Earth, Hello huge hot ball of strange matter! This was the source of the worries about the LHC, which did not eventuate since when the LHC was utilized no strangelets were noticeably produced.

One of the many unknowns about strangelets is their surface tension – nobody knows how to calculate this, at present. If the surface tension is strong enough, large stable strangelets should be possible – and potentially, strangelets with complex structure as femtotech requires.

And, of course, nobody knows what happens if you vibrate strangelets very, very fast with small amplitude – can you produce stable strangelets via dynamic stabilization? Could this be a path to viable femtotechnology, even if stable strangelets don't occur in nature? After all, carbon nanotubes appear not to occur in nature either.

The Future of Femtotech

So what's the bottom line – is there still more room at the bottom?

Nanotech is difficult engineering based on mostly known physics. Femtotech, on the other hand, pushes at the boundaries of known physics. When exploring possible routes to femtotech, one quickly runs up against cases where physicists just don't know the answer.

Degenerate matter of one form or another seems a promising potential route to femtotech. Bolonkin's speculations are intriguing, as are the possibilities of strangelets or novel weakly confined multi-quark systems. But the issue of stability is a serious one; nobody yet knows whether large strangelets can be made stable, or whether degenerate matter can be created at normal gravities, nor whether weakly confined quarks can be observed at normal temperatures, etc. Even where the relevant physics equations are believed to be known, the calculations are too hard to do given our present analytical and computational tools. And in some cases, e.g. strangelets, we run into situations where different physics theories held by respected physicists probably yield different answers.

String Theory

Putting my AI futurist hat on for a moment, I'm struck by what a wonderful example we have here of the potential for an only slightly superhuman AI to blast way past humanity in science and engineering. The human race seems on the verge of understanding particle physics well enough to analyze possible routes to femtotech. If a slightly superhuman AI, with a talent for physics, were to make a few small breakthroughs in computational physics, then it might (for instance) figure out how to make stable structures from degenerate matter at Earth gravity. Bolonkin-style femto-structures might then become plausible, resulting in femto-computing – and the slightly superhuman AI would then have a computational infrastructure capable of supporting massively superhuman AI. Can you say "singularity"? Of course, femtotech may be totally unnecessary in order for a Vingean singularity to occur (in fact I strongly suspect so). But be that as it may, it's interesting to think about just how much practical technological innovation might ensue from a relatively minor improvement in our understanding of fundamental physics.

Is it worth thinking about femtotech now, when the topic is wrapped up with so much unresolved physics? I think it is, if for no other reason than to give the physicists a nudge in certain directions that might otherwise be neglected. Most particle physics work – even experimental work with particle accelerators – seems to be motivated mainly by abstract theoretical interest. And there's nothing wrong with this – understanding the world is a laudable aim in itself; and

String Theory

furthermore, over the course of history, scientists aiming to understand the world have spawned an awful lot of practically useful by-products. But it's interesting to realize that there are potentially huge practical implications waiting in the wings, once particle physics advances a little more – if it advances in the right directions.

So, hey, all you particle physicists and physics funding agency program managers reading this article (and grumbling at my oversimplifications; sorry, this is tough stuff to write about for a nontechnical audience!), please take note – why not focus some attention on exploring the possibility of complexly structured degenerate matter under Earthly conditions, and other possibly femtotech-related phenomena such as those mentioned in Hugo de Garis's companion essay.

Is there still plenty more room at the bottom, after the nanoscale is fully explored? It seems quite possibly so – but we need to understand what goes on way down there a bit better before we can build stuff at the femtoscale. Fortunately, given the exponentially accelerating progress we're seeing in some relevant areas of technology, the wait for this understanding and the ensuing technologies may not be all that long.

There's lots of talk and research results lately concerning nanotech, i.e. molecular scale engineering and computation. In the winter of 90-91, I was at MIT listening to a nanotech question and answer session with a telephone hookup to Eric

String Theory

Drexler in California. I asked him if a femtotech was possible. The audience sniggered. I think Drexler said it would NOT be possible. I asked him again in early 93. Again he said it would NOT be possible. This reminded me of Rutherford's famous answer to the possibility of industrial application of nuclear energy, namely "moonshine", i.e. "rubbish". I thought it was ironic that the person who has devoted his life to propagandizing the coming of nanotech, would be so cynical about a possible femtotech. It struck me as an historical irony. The arguments he used (off the top of his head, I had the impression), were of the same style as those initially used against him when he was first propounding nanotech in the 70s, (e.g. molecular thermal motion would preclude it, Heisenberg's uncertainty principle would make it impossible, etc., etc.). In short, I felt that Drexler had not given the femtotech issue sufficient thought to be convincing. At least, I wasn't convinced.

What would be the long term implications if it is true that NO femtotech is possible? It would mean that humanity (or post humanity) would be limited to the nano level, i.e. there would always be a barrier to downscaling from the nano level to smaller scales, which would create a growing sense of frustration and discontent. A (sub)nuclear scale engineering and computation would do to nanotech what nanotech is now doing to microtech. It would allow a speed up of millions (over nanotech) and a density increase of quintillions (10 to power 18), i.e. a total performance increase (i.e. density times speed) of septillions (10 to power 24, or trillion of trillions). If a femtotech is possible, then I

String Theory

can imagine that ultra-intelligent civilizations out there (or right here) are of (sub)nuclear size. (Maybe elementary particles serve as the base for an otto-tech?). Leo Szilard, the guy who invented the idea of a nuclear chain reaction, and hence refuted Rutherford's skepticism, was cynical of Rutherford's conventional wisdom and set about trying to find a way for nuclear energy to be exploitable. I would like to do the same for femtotech. If it can be done, then how? What physical phenomena might be usable at a femto level?

I was visiting the Santa Fe Institute (SFI) in April/May of 1997, and had an opportunity to chat a bit with Murray Gell-Mann, the father of the quark and a Nobel Prize winner in particle physics, SFI's most famous member. I had been putting off writing this essay concerning the possibility of a "femto-tech", i.e. a technology based on femtosecond and femtometer phenomena. Since the femtoscale is nuclear and smaller, I needed to talk to a particle physicist. Murray seemed a good bet. I asked him "Can you think of any phenomenon which might serve as a basis for femto scale engineering and computing?" I told him that so far the only things I had come across that might be possible bases were:

    a) Nucleon Chemistry on the surface of neutron stars?
    b) b) Stranglets (agglomerations of S(trange) quarks)?

If strangelets are stable, as some people think, they could be of macro (humanly visible) size. Murray immediately surprised me with his reply which I duly wrote down. He said that some 25 years ago he had organized a meeting with

String Theory

some of his colleagues, "Most of whom are now dead", he said, "to look into the possible industrial applications of K(01) and K(02) kaons.

I didn't know what he was talking about, not being a particle physicist, but I looked up the kaons in my particle physics books at home and found them. What was not clear to me was how such particles might serve as a technological basis for a femtotech. I asked Murray if I could email him later for details. He said ok, so I did, but so far have not had a reply, which might have enabled me to give this femtotech essay more substance. Several nanotech magazines have been hounding me to get this essay written. I like pioneering new ideas, so I'll continue to push the femtotech thing. If some particle or nuclear physicists are reading this, and want to contact Murray or me on this issue, then his email is mgm@santafe.edu and mine is degaris@hip.atr.co.jp and my web is http://www.hip.atr.co.jp/~degaris. I don't have the necessary femto scale knowledge to be a competent pioneer in a possible femtotech, so I need help. Help!

End of this Internet Scientist's speculative articles.

String Theory

Chapter 9. Cosmological Implications of String Theory

String theory leads to some amazing (and controversial) implications. Although string theory is fascinating in its own right, what may prove to be even more intriguing are the possibilities that result from it.

Parallel Universes:

Some interpretations of string theory predict that our universe is not the only one. In fact, in the most extreme versions of the theory, an infinite number of other universes exist, some of which contain exact duplicates of our own universe.

As wild as this theory is, it's predicted by current research studying the very nature of the cosmos itself. In fact, parallel universes aren't just predicted by string theory — one view of quantum physics has suggested the theoretical existence of a certain type of parallel universe for more than half a century.

String Theory

Wormholes:

Einstein's theory of relativity predicts warped space called a *wormhole* (also called an *Einstein-Rosen bridge*). In this case, two distant regions of space are connected by a shorter wormhole, which gives a shortcut between those two distant regions, as shown in the figure below.

String theory allows for the possibility that wormholes extend not only between distant regions of our own universe, but also between distant regions of parallel universes. Perhaps universes that have different physical laws could even be connected by wormholes.

In fact, it's not clear whether wormholes will exist within string theory at all. As a quantum gravity theory, it's possible that the general relativity solutions that give rise to potential wormholes might go away.

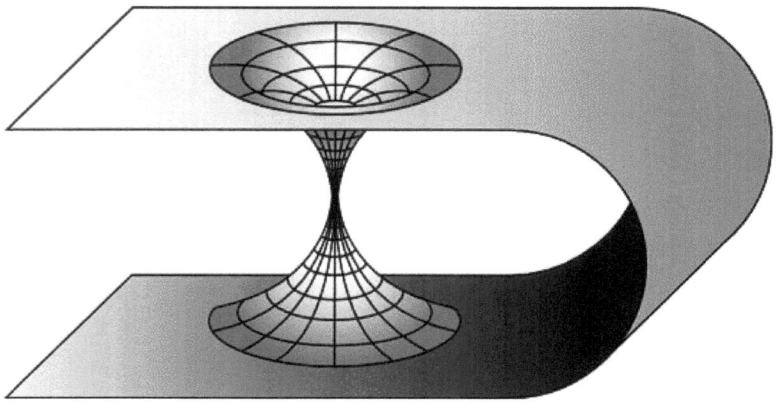

String Theory

The Universe as a Hologram

In the mid-1990s, two physicists came up with an idea called the *holographic principle*. In this theory, if you have a volume of space, you can take all the information contained in that space and show that it corresponds to information "written" on the surface of the space. As odd as it seems, this holographic principle may be key in resolving a major mystery of black holes that has existed for more than 20 years!

Many physicists believe that the holographic principle will be one of the fundamental physical principles that will allow insights into a greater understanding of string theory.

String Theory and Time Travel:

Some physicists believe that string theory may allow for multiple dimensions of time (by no means the dominant view). As our understanding of time grows with string theory, it's possible that scientists may discover new means of traveling through the time dimension or show that such theoretical possibilities are, in fact, impossible, as most physicists believe.

String Theory and the Big Bang

String theory is being applied to cosmology, which means that it may give us insights into the formation of the universe. The exact implications are still being explored, but some

String Theory

believe that string theory supports the current cosmological model of inflation, while others believe it allows for entirely universal creation scenarios.

Inflation theory predicts that, very shortly after the original big bang, the universe began to undergo a period of rapid, exponential inflation. This theory, which applies principles of particle physics to the early universe as a whole, is seen by many as the only way to explain some properties of the early universe.

In string theory, there also exists a possible alternate model to our current big bang model in which two branes collided together and our universe is the result. In this model, called the *ekpyrotic universe,* the universe goes through cycles of creation and destruction, over and over.

The End of the Universe:

The ultimate fate of the universe is a question that physics has long explored, and a final version of string theory may help us ultimately determine the matter density and cosmological constant of the universe. By determining these values, cosmologists will be able to determine whether our universe will ultimately contract in upon itself, ending in a big crunch -- and perhaps start all over again.

String Theory

Chapter 10. What are Strings *made of* in String Theory?

When you ask "what is something made of", you imply that this something is a composite system made of more "fundamental" or "elementary" components.

In most cases, this question would make sense. For example, a grain of sand is made of molecules. The molecules are made of atoms. The atoms are made of electrons, protons and neutrons. The protons and neutrons are made of quarks.

However, our current understanding is that elementary particles such as electrons and quarks are fundamental; they are not "made of" anything. Instead, they "make up" everything else!

Some physicists think that elementary particles such as electrons and quarks are made of even more fundamental things, called "strings". It's very important to stress that although this hypothesis is very intriguing and has led to the development of countless useful mathematical tools and models, it has not been proven experimentally.

To address your concerns in the question details, the strings in string theory are nothing at all like "normal" strings made

String Theory

of rope or some other material. They are called "strings" mostly for convenience. A more appropriate name would be perhaps "1-dimensional fundamental constituents", but "strings" is just more catchy

A very common question is: "are strings made of energy"? The answer is no. To explain, let's go back to elementary particles such as electrons, since they are more familiar to most people.

Are elementary particles made of energy? No, they have energy as one of their properties. You can measure a particle's energy. A particle also has other properties such as mass, momentum and spin. The particle is not "made of" mass, it's not "made of" momentum, it's not "made of" spin, and it's also not "made of" energy.

As an analogy from daily life, if you take a piece of metal for example, you can measure its properties, such as temperature. However, no one will ever claim that metal is "made of" temperature. This just doesn't make sense. Temperature is just a property of metal. Similarly, energy is a property of elementary particles. But it doesn't make sense to say that they are "made of" that energy.

If you are still not convinced, think about this: just like a piece of metal can be cooled to absolute zero (so it doesn't have any temperature), a particle can have no energy, at least in principle. So, what is a particle with no energy made of?

The exact same arguments also apply to strings.

String Theory

Addendum - on "classical" vs. "quantum" intuition and concepts:

I suspect that the problem most people have with coming to terms with the notion of a fundamental particle (or string, or whatever) is that the words "made of" have a certain common meaning from our daily life which just cannot be applied to physics at the quantum level.

The famous quote "no one understands quantum mechanics" means that we cannot understand quantum things using our "classical" intuition from our daily lives. But we understand it perfectly well if we use the new concepts and ideas introduced by quantum mechanics. And fundamental particles are one of these new concepts that we cannot understand with our old "classical" intuition alone.

The reason fundamental particles (or strings, if they actually exist) are called fundamental is that they are the point beyond which you cannot go any further to something "more fundamental". Let me illustrate this as follows:

What is a chair made of? Wood.

What is wood made of? Carbon and oxygen molecules (mostly).

What are molecules made of? Atoms.

What are atoms made of? Protons, neutrons and electrons.

String Theory

What are protons made of? Quarks.

What are quarks made of? Uh... Sorry, this question simply doesn't make sense because there is nothing more fundamental than quarks.

It's like asking "what is north of the north pole" or "what number is larger than infinity". These questions are impossible to answer because the questions themselves are meaningless.

Unless string theory is true, in which case quarks are made of strings, at least in some sense. But then the strings are the most fundamental thing, and asking what they are made of is again a meaningless question.

Appendix A. References and Bibliographies

Notes

1. **Jump up^** For example, physicists are still working to understand the phenomenon of quark confinement, the paradoxes of black holes, and the origin of dark energy.
2. **Jump up^** For example, in the context of the AdS/CFT correspondence, theorists often formulate and study theories of gravity in unphysical numbers of spacetime dimensions.
3. **Jump up^** *"Top Cited Articles during 2010 in hep-th"*. *Retrieved 25 July 2013*.
4. **Jump up^** More precisely, one cannot apply the methods of perturbative quantum field theory.
5. **Jump up^** Two independent mathematical proofs of mirror symmetry were given by Givental 1996, 1998 and Lian, Liu, Yau 1997, 1999, 2000.
6. **Jump up^** More precisely, a nontrivial group is called *simple* if its only normal subgroups are the trivial group and the group itself. The Jordan–Hölder theorem exhibits finite simple groups as the building blocks for all finite groups.

String Theory

Citations

1. ^ Jump up to:ᵃ ᵇ Becker, Becker, and Schwarz 2007, p. 1
2. **Jump up^** Zwiebach 2009, p. 6
3. ^ Jump up to:ᵃ ᵇ Becker, Becker, and Schwarz 2007, pp. 2–3
4. **Jump up^** Becker, Becker, and Schwarz 2007, pp. 9–12
5. **Jump up^** Becker, Becker, and Schwarz 2007, pp. 14–15
6. ^ Jump up to:ᵃ ᵇ Klebanov and Maldacena 2009
7. ^ Jump up to:ᵃ ᵇ Merali 2011
8. **Jump up^** Sachdev 2013
9. **Jump up^** Becker, Becker, and Schwarz 2007, pp. 3, 15–16
10. **Jump up^** Becker, Becker, and Schwarz 2007, p. 8
11. **Jump up^** Becker, Becker, and Schwarz 13–14
12. ^ Jump up to:ᵃ ᵇ Woit 2006
13. ^ Jump up to:ᵃ ᵇ Zee 2010
14. **Jump up^** Becker, Becker, and Schwarz 2007, p. 2
15. ^ Jump up to:ᵃ ᵇ Becker, Becker, and Schwarz 2007, p. 6
16. **Jump up^** Zwiebach 2009, p. 12
17. **Jump up^** Becker, Becker, and Schwarz 2007, p. 4
18. **Jump up^** Zwiebach 2009, p. 324
19. **Jump up^** Wald 1984, p. 4
20. **Jump up^** Zee 2010, Parts V and VI
21. **Jump up^** Zwiebach 2009, p. 9
22. **Jump up^** Zwiebach 2009, p. 8
23. ^ Jump up to:ᵃ ᵇ Yau and Nadis 2010, Ch. 6
24. **Jump up^** Greene 2000, p. 186
25. **Jump up^** Yau and Nadis 2010, p. ix
26. **Jump up^** Randall and Sundrum 1999
27. ^ Jump up to:ᵃ ᵇ Becker, Becker, and Schwarz 2007
28. **Jump up^** Zwiebach 2009, p. 376
29. ^ Jump up to:ᵃ ᵇ Moore 2005, p. 214
30. **Jump up^** Moore 2005, p. 215
31. ^ Jump up to:ᵃ ᵇ Aspinwall et al. 2009
32. ^ Jump up to:ᵃ ᵇ Kontsevich 1995
33. **Jump up^** Kapustin and Witten 2007
34. ^ Jump up to:ᵃ ᵇ Duff 1998

String Theory

35. **Jump up^** Duff 1998, p. 64
36. **Jump up^** Nahm 1978
37. **Jump up^** Cremmer, Julia, and Scherk 1978
38. ^ Jump up to:*a b c d e* Duff 1998, p. 65
39. **Jump up^** Sen 1994a
40. **Jump up^** Sen 1994b
41. **Jump up^** Hull and Townsend 1995
42. **Jump up^** Duff 1998, p. 67
43. **Jump up^** Bergshoeff, Sezgin, and Townsend 1987
44. **Jump up^** Duff et al. 1987
45. **Jump up^** Duff 1998, p. 66
46. **Jump up^** Witten 1995
47. **Jump up^** Duff 1998, pp. 67–68
48. **Jump up^** Becker, Becker, and Schwarz 2007, p. 296
49. **Jump up^** Hořava and Witten 1996
50. **Jump up^** Duff 1996, sec. 1
51. ^ Jump up to:*a b c* Banks et al. 1997
52. **Jump up^** Connes 1994
53. **Jump up^** Connes, Douglas, and Schwarz 1998
54. **Jump up^** Nekrasov and Schwarz 1998
55. **Jump up^** Seiberg and Witten 1999
56. ^ Jump up to:*a b c* de Haro et al. 2013, p. 2
57. **Jump up^** Yau and Nadis 2010, p. 187–188
58. **Jump up^** Bekenstein 1973
59. ^ Jump up to:*a b* Hawking 1975
60. **Jump up^** Wald 1984, p. 417
61. **Jump up^** Yau and Nadis 2010, p. 189
62. ^ Jump up to:*a b* Strominger and Vafa 1996
63. **Jump up^** Yau and Nadis 2010, pp. 190–192
64. **Jump up^** Maldacena, Strominger, and Witten 1997
65. **Jump up^** Ooguri, Strominger, and Vafa 2004
66. **Jump up^** Yau and Nadis 2010, pp. 192–193
67. **Jump up^** Yau and Nadis 2010, pp. 194–195
68. **Jump up^** Strominger 1998
69. **Jump up^** Guica et al. 2009
70. **Jump up^** Castro, Maloney, and Strominger 2010
71. ^ Jump up to:*a b* Maldacena 1998
72. ^ Jump up to:*a b* Gubser, Klebanov, and Polyakov 1998

String Theory

73. ^ Jump up to:a b Witten 1998
74. **Jump up^** Klebanov and Maldacena 2009, p. 28
75. ^ Jump up to:a b c Maldacena 2005, p. 60
76. ^ Jump up to:a b Maldacena 2005, p. 61
77. **Jump up^** Zwiebach 2009, p. 552
78. **Jump up^** Maldacena 2005, pp. 61–62
79. **Jump up^** Susskind 2008
80. **Jump up^** Zwiebach 2009, p. 554
81. **Jump up^** Maldacena 2005, p. 63
82. **Jump up^** Hawking 2005
83. **Jump up^** Zwiebach 2009, p. 559
84. ^ Jump up to:a b Kovtun, Son, and Starinets 2001
85. ^ Jump up to:a b Merali 2011, p. 303
86. **Jump up^** Luzum and Romatschke 2008
87. **Jump up^** Sachdev 2013, p. 51
88. **Jump up^** Candelas et al. 1985
89. **Jump up^** Yau and Nadis 2010, pp. 147–150
90. **Jump up^** Becker, Becker, and Schwarz 2007, pp. 530–531
91. **Jump up^** Becker, Becker, and Schwarz 2007, p. 531
92. **Jump up^** Becker, Becker, and Schwarz 2007, p. 538
93. **Jump up^** Becker, Becker, and Schwarz 2007, p. 533
94. **Jump up^** Becker, Becker, and Schwarz 2007, pp. 539–543
95. **Jump up^** Deligne et al. 1999, p. 1
96. **Jump up^** Hori et al. 2003, p. xvii
97. **Jump up^** Aspinwall et al. 2009, p. 13
98. **Jump up^** Hori et al. 2003
99. **Jump up^** Yau and Nadis 2010, p. 167
100. **Jump up^** Yau and Nadis 2010, p. 166
101. ^ Jump up to:a b Yau and Nadis 2010, p. 169
102. **Jump up^** Candelas et al. 1991
103. **Jump up^** Yau and Nadis 2010, p. 171
104. **Jump up^** Hori et al. 2003, p. xix
105. **Jump up^** Strominger, Yau, and Zaslow 1996
106. **Jump up^** Dummit and Foote 2004
107. **Jump up^** Dummit and Foote 2004, pp. 102–103
108. ^ Jump up to:a b Klarreich 2015

String Theory

109. **Jump up^** Gannon 2006, p. 2
110. **Jump up^** Gannon 2006, p. 4
111. **Jump up^** Conway and Norton 1979
112. **Jump up^** Gannon 2006, p. 5
113. **Jump up^** Gannon 2006, p. 8
114. **Jump up^** Borcherds 1992
115. **Jump up^** Frenkel, Lepowsky, and Meurman 1988
116. **Jump up^** Gannon 2006, p. 11
117. **Jump up^** Eguchi, Ooguri, and Tachikawa 2010
118. **Jump up^** Cheng, Duncan, and Harvey 2013
119. **Jump up^** Duncan, Griffin, and Ono 2015
120. **Jump up^** Witten 2007
121. **Jump up^** Woit 2006, pp. 240–242
122. ^ Jump up to:[a] [b] Woit 2006, p. 242
123. **Jump up^** Weinberg 1987
124. **Jump up^** Woit 2006, p. 243
125. **Jump up^** Susskind 2005
126. **Jump up^** Woit 2006, pp. 242–243
127. **Jump up^** Woit 2006, p. 240
128. **Jump up^** Woit 2006, p. 249
129. **Jump up^** Smolin 2006, p. 81
130. **Jump up^** Smolin 2006, p. 184
131. ^ Jump up to:[a] [b] Polchinski 2007
132. ^ Jump up to:[a] [b] Lee Smolin, April 2007:"Archived copy". Archived from the original on November 5, 2015. RetrievedDecember 31, 2015. Response to review of The Trouble with Physics by Joe Polchinski
133. **Jump up^** Penrose 2004, p. 1017
134. **Jump up^** Woit 2006, pp. 224–225
135. **Jump up^** Woit 2006, Ch. 16
136. **Jump up^** Woit 2006, p. 239
137. **Jump up^** Penrose 2004, p. 1018
138. **Jump up^** Penrose 2004, pp. 1019–1020
139. **Jump up^** Smolin 2006, p. 349
140. **Jump up^** Smolin 2006, Ch. 20
141. http://www.kurzweilai.net/femtotech-computing-at-the-femtometer-scale-using-quarks-and-gluons,

String Theory

Bibliography

- *Aspinwall, Paul; Bridgeland, Tom; Craw, Alastair; Douglas, Michael; Gross, Mark; Kapustin, Anton; Moore, Gregory; Segal, Graeme; Szendrői, Balázs; Wilson, P.M.H., eds. (2009). Dirichlet Branes and Mirror Symmetry. American Mathematical Society. ISBN 978-0-8218-3848-8.*
- *Banks, Tom; Fischler, Willy; Schenker, Stephen; Susskind, Leonard (1997). "M theory as a matrix model: A conjecture". Physical Review D 55 (8): 5112–5128. arXiv:hep-th/9610043.Bibcode:1997PhRvD..55.5112B. doi:10.1103/physrevd.55.5112.*
- *Becker, Katrin; Becker, Melanie; Schwarz, John (2007). String theory and M-theory: A modern introduction. Cambridge University Press.ISBN 978-0-521-86069-7.*
- *Bekenstein, Jacob (1973). "Black holes and entropy". Physical Review D 7 (8): 2333–2346. Bibcode:1973PhRvD...7.2333B.doi:10.1103/PhysRevD.7.2333.*
- *Bergshoeff, Eric; Sezgin, Ergin; Townsend, Paul (1987). "Supermembranes and eleven-dimensional supergravity". Physics Letters B 189 (1): 75–78. Bibcode:1987PhLB..189...75B.doi:10.1016/0370-2693(87)91272-X.*
- *Borcherds, Richard (1992). "Monstrous moonshine and Lie superalgebras". Inventiones Mathematicae 109 (1): 405–444.Bibcode:1992InMat.109..405B. doi:10.1007/BF01232032.*
- *Candelas, Philip; de la Ossa, Xenia; Green, Paul; Parks, Linda (1991). "A pair of Calabi–Yau manifolds as an exactly soluble superconformal field theory". Nuclear Physics B 359 (1): 21–74.Bibcode:1991NuPhB.359...21C. doi:10.1016/0550-3213(91)90292-6.*
- *Candelas, Philip; Horowitz, Gary; Strominger, Andrew; Witten, Edward (1985). "Vacuum configurations for superstrings". Nuclear Physics B 258: 46–74. Bibcode:1985NuPhB.258...46C.doi:10.1016/0550-3213(85)90602-9.*
- *Castro, Alejandra; Maloney, Alexander; Strominger, Andrew (2010). "Hidden conformal symmetry of the Kerr black hole". Physical Review D 82 (2). arXiv:1004.0996. Bibcode:2010PhRvD..82b4008C.doi:10.1103/PhysRevD.82.024008.*

String Theory

- Cheng, Miranda; Duncan, John; Harvey, Jeffrey (2013). "Umbral Moonshine". arXiv:1204.2779.
- Connes, Alain (1994). Noncommutative Geometry. Academic Press.ISBN 978-0-12-185860-5.
- Connes, Alain; Douglas, Michael; Schwarz, Albert (1998). "Noncommutative geometry and matrix theory". Journal of High Energy Physics. 19981 (2): 003. arXiv:hep-th/9711162.Bibcode:1998JHEP...02..003C. doi:10.1088/1126-6708/1998/02/003.
- Conway, John; Norton, Simon (1979). "Monstrous moonshine". Bull. London Math. Soc. **11** (3): 308–339. doi:10.1112/blms/11.3.308.
- Cremmer, Eugene; Julia, Bernard; Scherk, Joel (1978). "Supergravity theory in eleven dimensions". Physics Letters B **76** (4): 409–412.Bibcode:1978PhLB...76..409C. doi:10.1016/0370-2693(78)90894-8.
- de Haro, Sebastian; Dieks, Dennis; 't Hooft, Gerard; Verlinde, Erik (2013). "Forty Years of String Theory Reflecting on the Foundations". Foundations of Physics **43** (1): 1–7.Bibcode:2013FoPh...43....1D. doi:10.1007/s10701-012-9691-3.
- Deligne, Pierre; Etingof, Pavel; Freed, Daniel; Jeffery, Lisa; Kazhdan, David; Morgan, John; Morrison, David; Witten, Edward, eds. (1999). Quantum Fields and Strings: A Course for Mathematicians **1**. American Mathematical Society. ISBN 978-0821820124.
- Duff, Michael (1996). "M-theory (the theory formerly known as strings)". International Journal of Modern Physics A **11** (32): 6523–41. arXiv:hep-th/9608117. Bibcode:1996IJMPA..11.5623D.doi:10.1142/S0217751X96002583.
- Duff, Michael (1998). "The theory formerly known as strings".Scientific American **278** (2): 64–9.doi:10.1038/scientificamerican0298-64.
- Duff, Michael; Howe, Paul; Inami, Takeo; Stelle, Kellogg (1987). "Superstrings in $D=10$ from supermembranes in $D=11$". Nuclear Physics B **191** (1): 70–74. Bibcode:1987PhLB..191...70D.doi:10.1016/0370-2693(87)91323-2.
- Dummit, David; Foote, Richard (2004). Abstract Algebra. Wiley.ISBN 978-0-471-43334-7.

String Theory

- Duncan, John; Griffin, Michael; Ono, Ken (2015). "Proof of the Umbral Moonshine Conjecture". arXiv:1503.01472.
- Eguchi, Tohru; Ooguri, Hirosi; Tachikawa, Yuji (2011). "Notes on the K3 surface and the Mathieu group M_{24}". Experimental Mathematics **20** (1): 91–96. doi:10.1080/10586458.2011.544585.
- Frenkel, Igor; Lepowsky, James; Meurman, Arne (1988). Vertex Operator Algebras and the Monster. Pure and Applied Mathematics **134**. Academic Press. ISBN 0-12-267065-5.
- Gannon, Terry. Moonshine Beyond the Monster: The Bridge Connecting Algebra, Modular Forms, and Physics. Cambridge University Press.
- Givental, Alexander (1996). "Equivariant Gromov-Witten invariants".International Mathematics Research Notices **1996** (13): 613–663.doi:10.1155/S1073792896000414.
- Givental, Alexander (1998). "A mirror theorem for toric complete intersections". Topological field theory, primitive forms and related topics: 141–175. doi:10.1007/978-1-4612-0705-4_5. ISBN 978-1-4612-6874-1.
- Gubser, Steven; Klebanov, Igor; Polyakov, Alexander (1998). "Gauge theory correlators from non-critical string theory". Physics Letters B**428**: 105–114. arXiv:hep-th/9802109.Bibcode:1998PhLB..428..105G. doi:10.1016/S0370-2693(98)00377-3.
- Guica, Monica; Hartman, Thomas; Song, Wei; Strominger, Andrew (2009). "The Kerr/CFT Correspondence". Physical Review D **80** (12).arXiv:0809.4266. Bibcode:2009PhRvD..80l4008G.doi:10.1103/PhysRevD.80.124008.
- Hawking, Stephen (1975). "Particle creation by black holes".Communications in Mathematical Physics **43** (3): 199–220.Bibcode:1975CMaPh..43..199H. doi:10.1007/BF02345020.
- Hawking, Stephen (2005). "Information loss in black holes". Physical Review D **72** (8). arXiv:hep-th/0507171.Bibcode:2005PhRvD..72h4013H.doi:10.1103/PhysRevD.72.084013.
- Hořava, Petr; Witten, Edward (1996). "Heterotic and Type I string dynamics from eleven dimensions". Nuclear Physics B **460** (3): 506–524. arXiv:hep-th/9510209. Bibcode:1996NuPhB.460..506H.doi:10.1016/0550-3213(95)00621-4.
- Hori, Kentaro; Katz, Sheldon; Klemm, Albrecht; Pandharipande, Rahul; Thomas, Richard; Vafa, Cumrun; Vakil, Ravi; Zaslow,

- Eric, eds. (2003). *Mirror Symmetry* (PDF). American Mathematical Society. ISBN 0-8218-2955-6.
- Hull, Chris; Townsend, Paul (1995). "Unity of superstring dualities".*Nuclear Physics B* **4381** (1): 109–137. arXiv:hep-th/9410167.Bibcode:1995NuPhB.438..109H. doi:10.1016/0550-3213(94)00559-W.
- Kapustin, Anton; Witten, Edward (2007). "Electric-magnetic duality and the geometric Langlands program". *Communications in Number Theory and Physics* **1** (1): 1–236. arXiv:hep-th/0604151.Bibcode:2007CNTP....1....1K. doi:10.4310/cntp.2007.v1.n1.a1.
- Klarreich, Erica. "Mathematicians chase moonshine's shadow".*Quanta Magazine*. Retrieved March 2015.
- Klebanov, Igor; Maldacena, Juan (2009). "Solving Quantum Field Theories via Curved Spacetimes" (PDF). *Physics Today* **62**: 28–33. Bibcode:2009PhT....62a..28K. doi:10.1063/1.3074260. Archived from the original (PDF) on July 2, 2013. Retrieved May 2013.
- Kontsevich, Maxim (1995). "Homological algebra of mirror symmetry".*Proceedings of the International Congress of Mathematicians*: 120–139. arXiv:alg-geom/9411018. Bibcode:1994alg.geom.11018K.
- Kovtun, P. K.; Son, Dam T.; Starinets, A. O. (2001). "Viscosity in strongly interacting quantum field theories from black hole physics".*Physical Review Letters* **94** (11): 111601. arXiv:hep-th/0405231.Bibcode:2005PhRvL..94k1601K.doi:10.1103/PhysRevLett.94.111601. PMID 15903845.
- Lian, Bong; Liu, Kefeng; Yau, Shing-Tung (1997). "Mirror principle, I". *Asian Journal of Mathematics* **1**: 729–763. arXiv:alg-geom/9712011. Bibcode:1997alg.geom.12011L.
- Lian, Bong; Liu, Kefeng; Yau, Shing-Tung (1999a). "Mirror principle, II". *Asian Journal of Mathematics* **3**: 109–146.arXiv:math/9905006. Bibcode:1999math......5006L.
- Lian, Bong; Liu, Kefeng; Yau, Shing-Tung (1999b). "Mirror principle, III". *Asian Journal of Mathematics* **3**: 771–800.arXiv:math/9912038. Bibcode:1999math.....12038L.
- Lian, Bong; Liu, Kefeng; Yau, Shing-Tung (2000). "Mirror principle, IV". *Surveys in Differential Geometry* **7**: 475–496.arXiv:math/0007104. Bibcode:2000math......7104L.doi:10.4310/sdg.2002.v7.n1.a15.

String Theory

- Luzum, Matthew; Romatschke, Paul (2008). "Conformal relativistic viscous hydrodynamics: Applications to RHIC results at $\sqrt{s_{NN}}=200 GeV$". Physical Review C **78** (3). arXiv:0804.4015.doi:10.1103/PhysRevC.78.034915.
- Maldacena, Juan (1998). "The Large N limit of superconformal field theories and supergravity". Advances in Theoretical and Mathematical Physics **2**: 231–252. arXiv:hep-th/9711200.Bibcode:1998AdTMP...2..231M. doi:10.1063/1.59653.
- Maldacena, Juan (2005). "The Illusion of Gravity" (PDF). Scientific American **293** (5): 56–63. Bibcode:2005SciAm.293e..56M.doi:10.1038/scientificamerican1105-56. PMID 16318027. Archived from the original (PDF) on November 1, 2014. Retrieved July 2013.
- Maldacena, Juan; Strominger, Andrew; Witten, Edward (1997). "Black hole entropy in M-theory". Journal of High Energy Physics **1997** (12). doi:10.1088/1126-6708/1997/12/002.
- Merali, Zeeya (2011). "Collaborative physics: string theory finds a bench mate". Nature **478** (7369): 302–304.Bibcode:2011Natur.478..302M. doi:10.1038/478302a.PMID 22012369.
- Moore, Gregory (2005). "What is ... a Brane?" (PDF). Notices of the AMS **52**: 214. Retrieved June 2013.
- Nahm, Walter (1978). "Supersymmetries and their representations".Nuclear Physics B **135** (1): 149–166.Bibcode:1978NuPhB.135..149N. doi:10.1016/0550-3213(78)90218-3.
- Nekrasov, Nikita; Schwarz, Albert (1998). "Instantons on noncommutative R^4 and (2,0) superconformal six dimensional theory". Communications in Mathematical Physics **198** (3): 689–703.arXiv:hep-th/9802068. Bibcode:1998CMaPh.198..689N.doi:10.1007/s002200050490.
- Ooguri, Hirosi; Strominger, Andrew; Vafa, Cumrun (2004). "Black hole attractors and the topological string". Physical Review D **70**(10). doi:10.1103/physrevd.70.106007.
- Polchinski, Joseph (2007). "All Strung Out?". American Scientist. Retrieved April 2015.
- Penrose, Roger (2005). The Road to Reality: A Complete Guide to the Laws of the Universe. Knopf. ISBN 0-679-45443-8.
- Randall, Lisa; Sundrum, Raman (1999). "An alternative to compactification". Physical Review Letters **83** (23): 4690–

4693. *arXiv*:*hep-th/9906064*. *Bibcode*:*1999PhRvL..83.4690R*. *doi*:*10.1103/PhysRevLett.83.4690*.
- Sachdev, Subir (2013). "Strange and stringy". Scientific American **308** (44): 44–51. *Bibcode*:*2012SciAm.308a..44S*. *doi*:*10.1038/scientificamerican0113-44*.
- Seiberg, Nathan; Witten, Edward (1999). "String Theory and Noncommutative Geometry". Journal of High Energy Physics **1999** (9): 032. *arXiv*:*hep-th/9908142*. *Bibcode*:*1999JHEP...09..032S*. *doi*:*10.1088/1126-6708/1999/09/032*.
- Sen, Ashoke (1994a). "Strong-weak coupling duality in four-dimensional string theory". International Journal of Modern Physics A **9** (21): 3707–3750. *arXiv*:*hep-th/9402002*. *Bibcode*:*1994IJMPA...9.3707S*. *doi*:*10.1142/S0217751X94001497*.
- Sen, Ashoke (1994b). "Dyon-monopole bound states, self-dual harmonic forms on the multi-monopole moduli space, and $SL(2,\mathbb{Z})$ invariance in string theory". Physics Letters B **329** (2): 217–221. *arXiv*:*hep-th/9402032*. *Bibcode*:*1994PhLB..329..217S*. *doi*:*10.1016/0370-2693(94)90763-3*.
- Smolin, Lee (2006). The Trouble with Physics: The Rise of String Theory, the Fall of a Science, and What Comes Next. New York: Houghton Mifflin Co. ISBN 0-618-55105-0.
- Strominger, Andrew (1998). "Black hole entropy from near-horizon microstates". Journal of High Energy Physics **1998** (2): 009. *arXiv*:*hep-th/9712251*. *Bibcode*:*1998JHEP...02..009S*. *doi*:*10.1088/1126-6708/1998/02/009*.
- Strominger, Andrew; Vafa, Cumrun (1996). "Microscopic origin of the Bekenstein–Hawking entropy". Physics Letters B **379** (1): 99–104. *arXiv*:*hep-th/9601029*. *Bibcode*:*1996PhLB..379...99S*. *doi*:*10.1016/0370-2693(96)00345-0*.
- Strominger, Andrew; Yau, Shing-Tung; Zaslow, Eric (1996). "Mirror symmetry is T-duality". Nuclear Physics B **479** (1): 243–259. *arXiv*:*hep-th/9606040*. *Bibcode*:*1996NuPhB.479..243S*. *doi*:*10.1016/0550-3213(96)00434-8*.

String Theory

- *Susskind, Leonard (2005). The Cosmic Landscape: String Theory and the Illusion of Intelligent Design. Back Bay Books. ISBN 978-0316013338.*
- *Susskind, Leonard (2008). The Black Hole War: My Battle with Stephen Hawking to Make the World Safe for Quantum Mechanics. Little, Brown and Company. ISBN 978-0-316-01641-4.*
- *Wald, Robert (1984). General Relativity. University of Chicago Press. ISBN 978-0-226-87033-5.*
- *Weinberg, Steven (1987). Anthropic bound on the cosmological constant **59**. Physical Review Letters. p. 2607.*
- *Witten, Edward (1995). "String theory dynamics in various dimensions". Nuclear Physics B **443** (1): 85–126. arXiv:hep-th/9503124. Bibcode:1995NuPhB.443...85W. doi:10.1016/0550-3213(95)00158-O.*
- *Witten, Edward (1998). "Anti-de Sitter space and holography".Advances in Theoretical and Mathematical Physics **2**: 253–291.arXiv:hep-th/9802150. Bibcode:1998AdTMP...2..253W.*
- *Witten, Edward (2007). "Three-dimensional gravity revisited".arXiv:0706.3359 [hep-th].*
- *Woit, Peter (2006). Not Even Wrong: The Failure of String Theory and the Search for Unity in Physical Law. Basic Books. p. 105.ISBN 0-465-09275-6.*
- *Yau, Shing-Tung; Nadis, Steve (2010). The Shape of Inner Space: String Theory and the Geometry of the Universe's Hidden Dimensions. Basic Books. ISBN 978-0-465-02023-2.*
- *Zee, Anthony (2010). Quantum Field Theory in a Nutshell (2nd ed.). Princeton University Press. ISBN 978-0-691-14034-6.*
- *Zwiebach, Barton (2009). A First Course in String Theory. Cambridge University Press. ISBN 978-0-5*

http://www.kurzweilai.net/femtotech-computing-at-the-femtometer-scale-using-quarks-and-gluons

String Theory

String Theory

www.ingramcontent.com/pod-product-compliance
Lightning Source LLC
Chambersburg PA
CBHW041241200526
45159CB00028B/26